Facing Climate Change Together

The vast majority of climate scientists now agree that human-induced climate change is a reality. However, there is much ongoing research and debate over the magnitude, rate, and regional distribution of change and the level of confidence assigned to future climate projections. Nevertheless, our global society is confronted with the urgent and immediate need for a wise response to potential climate change.

This volume brings together scientists from the United States and France to review the state of the art in climate change science; all of them have extensive experience with climate research and international collaboration. Scientific jargon has been minimized for readers of different backgrounds.

Each chapter provides a description of a particular aspect of the climate problem, its role in current climate change, the impacts that are occurring or may occur in the future, and its societal importance. The introductory chapter is a synthesis of the major findings presented in the book and draws from the most recent assessment reports of the Intergovernmental Panel on Climate Change (IPCC).

This book is written for scientists and students in a wide range of fields, such as atmospheric science, physics, chemistry, biology, geography, geology, and socioeconomics, who are not necessarily specialists in climatology but are seeking an accessible and broad review of climate change issues.

CATHERINE GAUTIER focuses her research in climate change science on the modeling and observation of cloud and aerosol effects on climate. She is also involved in educational aspects of climate change science and policy. Gautier has been a professor in the Geography Department of the University of California, Santa Barbara, since 1990, and directed the Institute for Computational Earth System Science from 1996 to 2002.

JEAN-LOUIS FELLOUS joined the French space agency, CNES, in 1982 as Program Manager of the US–French Topex/Poseidon oceanography satellite. He headed the Earth Observation Programs at CNES from 1998 to 2001, and was director for Ocean Research at Ifremer, the French ocean research institute, from 2001 to 2005. Fellous then worked with the European Space Agency as coordinator of Earth observation satellite programs related to climate, and was Executive Officer of the Committee on Earth Observation Satellites (CEOS). He is a co-president of the Joint Technical Commission on Oceanography and Marine Meteorology (JCOMM). He now works as Executive Director of the Committee on Space Research (COSPAR).

Facing Climate Change Together

CATHERINE GAUTIER
AND
JEAN-LOUIS FELLOUS

CAMBRIDGE
UNIVERSITY PRESS

CAMBRIDGE UNIVERSITY PRESS
Cambridge, New York, Melbourne, Madrid, Cape Town, Singapore, São Paulo, Delhi

Cambridge University Press
The Edinburgh Building, Cambridge CB2 8RU, UK

Published in the United States of America by Cambridge University Press, New York

www.cambridge.org
Information on this title: www.cambridge.org/9780521896825

© Odile Jacob Paris, 2008

This publication is in copyright. Subject to statutory exception
and to the provisions of relevant collective licensing agreements,
no reproduction of any part may take place without
the written permission of Cambridge University Press.

First published 2008

Printed in the United Kingdom at the University Press, Cambridge

A catalog record for this publication is available from the British Library.

Library of Congress Cataloging in Publication data

Facing climate change together / [edited by] Catherine Gautier and Jean-Louis Fellous ; foreword by D. James Baker.
    p. cm.
Includes bibliographical references and index.
ISBN 978-0-521-89682-5 (hardback)
1. Climatic changes – Environmental aspects. 2. Hydrologic cycle. 3. Global temperature changes. 4. Atmospheric chemistry – Environmental aspects. 5. Nature – Effect of human beings on. I. Gautier, Catherine, 1947– II. Fellous, J. L. (Jean-Louis)
QC981.8.C5F329   2008
551.6 – dc22   2007048067

ISBN 978-0-521-89682-5 hardback

Cambridge University Press has no responsibility for the persistence or
accuracy of URLs for external or third-party Internet Web sites referred to
in this publication, and does not guarantee that any content on such
Web sites is, or will remain, accurate or appropriate.

To my daughters Kristen and Julie
C.G.

In memoriam
Gérard David Fellous and Lucia Jean Strawson
31 March 1972, Propriano, Corsica

To the memory of Yoram J. Kaufman
Our friend and colleague

Contents

Contributors ix
Foreword by D. James Baker xi
Acknowledgments xv

Introduction 1
CATHERINE GAUTIER AND JEAN-LOUIS FELLOUS

1 The global consensus and the Intergovernmental Panel on Climate Change 12
RICHARD C. J. SOMERVILLE AND JEAN JOUZEL

2 Greenhouse effect, radiation budget, and clouds 30
CATHERINE GAUTIER AND HERVÉ LE TREUT

3 Atmospheric aerosols and climate change 49
YORAM J. KAUFMAN AND OLIVIER BOUCHER

4 The global water cycle and climate 62
MOUSTAFA T. CHAHINE AND PIERRE MOREL

5 Climate change effects on the water cycle over land 74
ERIC F. WOOD AND KATIA LAVAL

6 Ocean and climate 91
CARL WUNSCH AND JEAN-FRANÇOIS MINSTER

7 Ice and climate 110
RAYMOND C. SMITH AND FRÉDÉRIQUE RÉMY

8 The changing global carbon cycle: from the holocene to the anthropocene   121
   BERRIEN MOORE III AND PHILIPPE CIAIS

9 Challenges of climate change: an Arctic perspective and its implications   149
   ROBERT W. CORELL

10 On some impacts of climate change over Europe and the Atlantic   159
    JEAN-CLAUDE ANDRÉ

11 Atmospheric chemistry and climate interactions   176
    GUY BRASSEUR AND MARIE-LISE CHANIN

12 Observing system for climate   189
    HELEN M. WOOD AND JEAN-LOUIS FELLOUS

13 Climate and society: what is the human dimension?   204
    ROBERTA BALSTAD AND JEAN-CHARLES HOURCADE

Conclusions   217
S. ICHTIAQUE RASOOL AND JEAN-CLAUDE DUPLESSY

Glossary   233
References   243
Index   253
The color plates are between pages 176 and 177.

Contributors

Jean-Claude André
CERFACS, Toulouse, France

Roberta Balstad
CIESIN, Columbia University (LDEO), Palisades, NY, USA

Olivier Boucher
UK Met Office, Exeter, UK

Guy Brasseur
National Center for Atmospheric Research, Boulder, CO, USA

Moustafa T. Chahine
NASA/Jet Propulsion Laboratory, Pasadena, CA, USA

Marie-Lise Chanin
Service d'Aéronomie du CNRS, Verrières-le-Buisson, France

Philippe Ciais
Laboratoire des Sciences du Climat et de l'Environnement, Saclay, France

Robert W. Corell
The H. John Heinz III Center for Science, Economics and the Environment; and the American Meteorological Society, Washington, DC, USA

Jean-Claude Duplessy
Laboratoire des Sciences du Climat et de l'Environnement, CEA-CNRS, Gif-sur-Yvette, France

Jean-Louis Fellous
Committee on Space Research (COSPAR), Paris, France

Catherine Gautier
University of California at Santa Barbara, CA, USA

Contributors

Jean-Charles Hourcade
CIRED, Nogent-sur-Marne, France

Jean Jouzel
Institut Pierre Simon Laplace, Paris, France

Yoram Kaufman
NASA Goddard Space Flight Center, College Park, MD, USA

Katia Laval
Laboratoire de Météorologie Dynamique CNRS; and Université Pierre et Marie Curie Paris, France

Hervé Le Treut
Laboratoire de Météorologie Dynamique, Palaiseau, France

Jean-François Minster
Total, Paris, France

Berrien Moore III
Institute for the Study of Earth, Oceans and Space, University of New Hampshire Durham, NH, USA

Pierre Morel
University of Paris (Emeritus)

S. Ichtiaque Rasool
International Consultant (formerly for NASA), Paris/Washington, DC

Frédérique Rémy
LEGOS, Toulouse, France

Raymond C. Smith
University of California at Santa Barbara, CA, USA (Emeritus)

Richard C. J. Somerville
Scripps Institution of Oceanography, La Jolla, CA, USA (Emeritus)

Eric F. Wood
Princeton University, Princeton, NJ, USA

Helen M. Wood
NOAA, Silver Springs, MD, USA

Carl Wunsch
Massachusetts Institute of Technology, Cambridge, MA, USA

Foreword

D. JAMES BAKER

There's an old saying: 'Climate is what you expect; weather is what you get.' Or, according to a student quoted on the website of the University Corporation for Atmospheric Research: 'Climate tells you what clothes to buy, but weather tells you what clothes to wear.' But now, as humans are increasingly affecting the climate with increased energy use, this distinction is being blurred – we expected one climate and may well be getting another that is outside our experience. We're buying clothes, or more broadly setting policies, that may not be appropriate for a globally warmed world.

But exactly what will the climate be, and what specifically should we be doing about it? The problem is that some aspects of climate change are certain, and some aspects are uncertain. There are things in the science of climate that we know well, things we don't know, and some things that we may never know precisely. We do know that adding more carbon dioxide to the atmosphere will cause warming – but the amount and, hence the impact, of that warming is uncertain. The impact could range from small to catastrophic. If society is to deal with global climate change in all its complexity, citizens and policy makers alike will have to learn how to deal with what we know and what we don't know.

Facing Climate Change Together helps us do that. It brings a welcome addition to climate literature by filling a niche for those who want to see the true state of the art. The authors are all experts in their fields, the science is comprehensive and up-to-date, and the full range of subjects in the complex climate field is covered. This book will stand out as an excellent reference for the science of climate change. It is

> **D. James Baker** is an oceanographer, now working as a consultant to the Intergovernmental Oceanographic Commission of UNESCO and to the H. John Heinz III Center for Science, Economics and the Environment, and as a Senior Fellow at the London School of Economics. Previously, he was Administrator of the U.S. National Oceanic and Atmospheric Administration (NOAA) and President of the Academy of Natural Sciences of Philadelphia.

particularly important that the authors of the various chapters have been willing to take on uncertainties, as well as certainties. By facing up to the realities of what we know and don't know, the authors provide an excellent antidote to the climate skeptics who pick and choose information to make unfounded statements.

There is a curve of probabilities – a probability density function – that summarizes what we know, both from the physical models and from the study of historical data. The curve, reproduced in the book as Figure 10.7, shows that the most likely effect of doubling CO_2 would lead to a global temperature change of 2–3 °C. But, the curve also shows that there is some (lower) probability that the changes could be smaller – say, just 1°, or larger, say, 4 or more degrees. A warming of 1° or less is very different from a scientific and policy viewpoint than even 3°. At 1° or less, all the effects would take longer to manifest themselves and hence would be easier to deal with. At the expected and most likely value of 2–3°, there will be significant changes facing society. At the high end, a warming of 4–5° or more could be a disaster for Earth – at least for humans and much of the rest of life. Major disruptions would take place, causing social and economic upheaval. Prudent planning puts us in the middle of the curve, while at the same time following the maxim, "Hope for the best, prepare for the worst."

How can we either reduce the uncertainties or learn what parts of the system are most sensitive? This is one of the major themes of the book. In their introduction, Catherine Gautier and Jean-Louis Fellous make an important point:

> One has to recognize that the first research results in a new interdisciplinary field, when all components are initially connected, tend to increase uncertainties. Once the interconnections are better understood and eventually modeled, then improvement occurs and uncertainties are reduced. The inherent, chaotic nature of the climate system, however, ensures that some uncertainty will always remain.

Throughout the book, there is much discussion about the uncertainties because that is where the exciting science is being done. Each chapter gives a careful review of the science of one part of the climate system, carefully goes into what is known and what is not known, and emphasizes what research needs to be done.

To produce a book like this – with scientists from just two countries – is an innovative and interesting idea, and the US–French collaborators have come up with a comprehensive summary of the issues. In fact, the scope is global, and the research represented here reflects the scientific debate that is carried out by scientists in many countries. Scientists will have different views on the scientific aspects of the issues, but these differences are science-based, not country-based.

The big difference is in the way that science is viewed by the public and policy makers in each country. In the United States, there is a lower level of public comprehension of science issues in general, and the climate and related energy issues in particular, than in France. The US Congress has refused to take regulatory action on climate issues, from ratifying the Kyoto Treaty to establishing caps on carbon emissions. Any policy changes in the United States will have to overcome this barrier. France and Europe overall have been much more responsive to the need for action and are a model for what the rest of the world needs to do. The European Commission has developed a number of pan-European programs that are addressing both scientific and policy issues for climate change.

In the end, society will have the policies it needs only if all of its sectors are educated about the certainties, the uncertainties, and the range of sensible policy options for global climate change. The editors and authors of this book can be proud of the contribution they have made.

Acknowledgments

We are grateful to D. James Baker and Jacques Merle, who extensively reviewed the whole manuscript; to Tom Karl, Marc Jamous, and Marianne Magnani, who read part of it and provided useful comments; and to David Herring, Jean and Daphne Kaufman, and Lorraine Remer, who helped complete Chapter 3 after Yoram Kaufman's death. We particularly wish to acknowledge the talent and skill of Michel Grégoire, who produced the set of figures for both the English and French[1] editions of this book.

---

[1] *Comprendre le changement climatique*, sous la direction de Jean-Louis Fellous et Catherine Gautier, Editions Odile Jacob Sciences, 298 pp., Paris, 2007.

Introduction

CATHERINE GAUTIER AND JEAN-LOUIS FELLOUS

Like the canary in the coal mine, the climate changes already evident in the Arctic are a call to action.

—Susan Collins

L'empire du climat est le premier de tous les empires, parce qu'il forme la différence des caractères et des passions des hommes.

—Montesquieu, *L'esprit des Lois*

Why this book?

Confusing climate information

There is a broad consensus on the basics of climate change and a lot of debate about scientific uncertainties. For instance, we know that CO_2 absorbs radiation, CO_2 is increasing because of human activities, global temperature is increasing, Arctic temperature is increasing at a much faster rate, sea level is

---

Catherine Gautier's research in climate change science focuses on the modeling and observation of cloud and aerosol effects on climate, with an emphasis on absorbing aerosols (mineral and soot). She is also working on educational aspects of climate change science and policy. She has been a professor in the Geography Department of the University of California, Santa Barbara, since 1990, and directed the Institute for Computational Earth System Science from 1996 to 2002.

Jean-Louis Fellous worked as an atmospheric scientist before joining the French Space Agency, CNES, in 1982 as Program Manager of the US–French Topex/Poseidon oceanography satellite. He headed the Earth Observation Programs at CNES from 1998 to 2001, and was director for Ocean Research at Ifremer, the French ocean research institute, from 2001 to 2005. He presently works with the European Space Agency as coordinator of Earth observation satellite programs related to climate, environment, and security. He is the Executive Officer of the Committee on Earth Observation Satellites (CEOS) and a co-president of the Joint Technical Commission on Oceanography and Marine Meteorology (JCOMM).

rising, Arctic sea ice and most of its glaciers are melting, and the ocean is getting more acidic. This agreement on the basics is as important as the research required on the uncertainties. In this book, the reader will see much discussion about the uncertainties, because that is where the exciting science is being done. From a policy maker's point of view, the uncertainty can lead to lack of action. Yet it is important to realize that the need for action recommended in this book is, in fact, based on science that is reasonably certain.

The information on climate change that the public receives comes from multiple channels and can be confusing: meteorology versus climate, natural variability versus human-induced changes, historical versus current climate variations, global averages versus extreme events, and so forth.

The media, in its efforts to provide fair and balanced views, often presents the public with conflicting opinions when only few actually exist. This style of news presentation results in an excessive emphasis on contrarians' scientific perspectives. Fiction books written by novelists, such as Michael Crichton's *State of Fear*, are presented as scientifically valid documents, and their pretend science content tends to receive much more attention than it deserves. The climate change conversation has been influenced by both political and industrial lobbies, particularly the energy sector. In its early stage, the US climate debate has been unnecessarily polarized along party lines, thus preventing candid discussions that are crucial for advancing this complex and evolving science.

On the other hand, both the government and the public benefit from up-to-date, first-rate scientific information offered through a reporting procedure aimed at providing the best available scientific knowledge on climate, in support of the United Nations Framework Convention for Climate Change (UNFCCC). Providing policy-relevant information is the mandate of the Intergovernmental Panel on Climate Change (IPCC; see Chapter 1). *IPCC Assessment Reports* representing the scientific community consensus are published at four- to six-year intervals and are subject to a rigorous and lengthy scientific review process involving hundreds of experts.

Despite broad international scientific consensus, governments throughout the world adopt different attitudes toward climate change. In particular, the United States and a few other countries on one side, and many European and other countries on the other side, have different positions on the type of global action proposed in the Kyoto Protocol. The differing stances held by nations with similar levels of economic development present another source of confusion for the public. Some nations try to impress on others that the uncertainties in our understanding of climate change are still large enough to postpone action or that technological solutions might be sufficient in the future to reverse trends and avoid future critical situations. An urgent step for those governments still skeptical about taking action to prevent climate change on economic grounds is, therefore, to acknowledge the existence of a climate crisis that needs to be dealt with now.

The climate science community: consensus or divergence?

Basically, climate science consists of understanding the complex interacting climate system, building models and predicting climate, observing climate trends both in the present and in the distant past, and studying the impacts of climate change on society. Climate science, like all sciences, progresses through data-based and model-based analysis and the argumentation and debate of hypotheses raised by these data and models. Discussion and critical assessment are the rule. Despite a broad consensus within the scientific community on the reality and nature of climate change, there are some disagreements with regard to the multiple causes of these changes, the rates of changes, and the model projections of future changes. The uncertainty inherent in some data sets (e.g. paleo records, satellite short time series) can lead to a variety of explanations and is exacerbated by the inaccurate and questionable data sometimes used in climate change studies (see Chapter 11). These limited data sets often require clever interpretations to extract the appropriate climate information.

One should recognize that the climate problem is very complex and requires collaboration among different scientific disciplines in order to be understood (see Figure 1.1). Solving the climate change conundrum is far outside the reach of any individual scientist because it involves a multitude of processes interacting over diverse time- and space scales. Enormous and ever-increasing amounts of data have to be collected and analyzed, and the intrinsic variability of the climate system demands sophisticated instrumentation. Furthermore, climate scales necessitate long time series of well-calibrated observations. Complex climate models are required to both integrate data and produce predictions of future climate. The unambiguous attribution of climate change in general to either natural or human causes, or a combination of the two, is always difficult. Attribution of any particular event, such as hurricanes or heat waves, to climate change can be achieved only in the form of a probability, by comparison to the statistical norm (see Chapter 9).

Conflicting interpretations and varying results derive from measurements taken by different instruments examining different aspects of the same process (i.e. measuring the rainfall directly within a cloud versus the temperature of the top of the cloud when assessing precipitation amounts) or model computations made in distinct ways (e.g. using various assumptions about when a modeled cloud forms). These difficulties must be seen as an indication of the climate system's complexity and not as an indication of disagreement among scientists.

It is not surprising that dissimilar scientific opinions within the climate science community exist in both the United States and many European countries. There is no more divergence, however, between scientists on opposite sides of the Atlantic Ocean than there is among scientists on just one side. An overwhelming majority of scientists around the world are of the same opinion that Earth's climate is changing and that the rate of change of several key climate parameters is increasing. Furthermore, these scientists agree that the cause of this change can undoubtedly be attributed to the scale of fossil fuels use and, to a lesser extent, land-use changes and deforestation dating back to at least the beginning of the industrial era, as well as to natural causes and variability.

The overall message, therefore, is one of basic agreement within the scientific community, which acknowledges that human-induced climate change is real. In addition, most scientists concur that the situation is urgent and represents a major risk to our civilization and requires an immediate, coordinated response from all nations: this response must happen *now*.

Not just another book on climate change

Most chapters in this book have been written by a pair of world-class French and US experts. All of these authors have extensive experience with climate research and international collaboration. In writing their chapters, the authors have been challenged to get as close as possible to the verifiable 'truth' and to present this knowledge to the readers, informing them of the complexities of climate change that, in the end, may involve the survival of humanity on Earth. As a result, these scientists speak the strict, and often difficult, language of science: physics, chemistry, biology, geography, geology, and socioeconomic research. While communicating to the reader, they have attempted to avoid the scientific jargon of their discipline and use a well-codified vocabulary to minimize any misunderstanding and misinterpretation. Last but not least, the book is published in both countries and languages!

Many of these authors have collaborated on international programs[1]; some have studied in the same laboratories, and most have worked together in both the United States and Europe. As the reader will see, there is no unanimity among the author's opinions throughout the book, and one might even identify some differences of scientific points of view among the chapters. In some instances, divergent viewpoints between authors of the same chapter were either resolved through debate or remain 'between the lines' in the text.

---

[1] One of the Chapter 3 authors, Yoram Kaufman, passed away in May 2006 after a tragic accident. We offer a tribute to his enthusiastic participation in this collective endeavor. Yoram was both a colleague and a friend to many of us in the United States and in France.

This book is a fact-based analysis of climate change that is grounded in evidence. It 'talks' science: how the tiny particles that humans produce through their activities have the potential to change clouds and rainfall; how the gas we put in the atmosphere changes the water cycle, influencing El Niño events and hurricanes; how the use of our land affects the water that falls on it and its ability to absorb the greenhouse gases our activities put in the atmosphere; and so forth.

The first chapter provides a synthesis of the major findings presented in this book, and draws from the most recent outcome of the IPCC, the *Fourth Assessment Report* released in 2007. The last chapter presents views on climate and society as expressed by two prominent social scientists. One should recognize that, in addition to the description of the physical mechanisms involved, climate change is caused by social and economic development, has many social and economic consequences, and can only be dealt with through social and economic solutions, including technological ones, when they exist.

Each chapter of this book is intended to provide the reader with a description of a particular aspect of the climate problem, its role in current climate change, the impacts and feedbacks that are occurring or may occur in the future, and its societal importance. Together, the introduction and conclusion provide the anchoring axis around which the chapters are tied in a book that is, by necessity, a collection of essays that are both self-sufficient and interconnected.[2]

Cross-cutting themes

Four main themes are threaded throughout the book: the complexity, interconnectedness, and the uncertainty of climate change, and its impact on humans. *Complexity* arises from the nonlinear nature of a climate system that can evolve chaotically, with abrupt changes that are difficult to predict. Knowing the response of the climate system to a perturbation is not sufficient to derive its future behavior in response to larger, even qualitatively similar, perturbations because some thresholds (sometimes referred to as 'tipping points') may have been exceeded, and the system may then move to another state of equilibrium. *Interconnectedness* reflects the diversity of Earth's climate system that encompasses all components of the planet: hydrosphere, cryosphere, atmosphere, and biosphere. The interactions of these components are complicated and often nonlinear as well, acting over inherently different temporal and spatial scales and creating feedbacks whose magnitudes also are often difficult to predict. *Uncertainty* is a

---

[2] We would like to reassure the reader that this book, written with so many globally dispersed authors, genuinely takes on climate change: it is carbon free! Not one author traveled to do his or her work, although the writing and revisions did travel – through cyberspace, that is. Besides email, all communication among the participants occurred by teleconference.

central theme in climate science, and its reduction is a primary goal of scientists. Important climate uncertainties include climate sensitivity, forcings, and feedbacks. *Reducing Uncertainties* was the title of a brochure edited in 1992 by the International Geosphere–Biosphere Programme, a major scientific undertaking initiated in 1986 to study global change. One has to recognize that the first research results in a new interdisciplinary field, when all components are initially connected, tend to increase uncertainties. Once the interconnections are better understood and eventually modeled, improvement occurs and uncertainties are reduced. The inherent, chaotic nature of the climate system, however, ensures that some uncertainty will always remain. Finally, *human impact* is at the heart of our concerns, and the whole book is an attempt to highlight the urgent need for a sound human response to the threatening challenges ahead.

Models

To understand climate, scientists cannot perform laboratory experiments in which factors are controlled to test the sensitivity of the system under various conditions. It is impossible to conduct such full-size experiments, although human species may inadvertently be doing so through careless modification of the atmospheric composition, irresponsible use of land surfaces, and causing negative impacts on climate in general. Scientists have, however, the ability to perform virtual experiments through numerical models, a unique tool used to understand how the climate system behaves and to predict how it evolves.

For the nonspecialist, a model may seem like a 'black box' from which predictions about the future are dispensed. But, in reality, models – particularly climate models – are sophisticated mathematical computer programs that compute the state of the (climate) system at every instant based on physical laws, such as the conservation of mass, energy, and momentum, many of which were established centuries ago. The same fluid mechanics theory that realistically represents such complex flows as propellants in rocket motors or airflow around aisles of stratospheric aircraft works just as well when applied to climate modeling. In climate models, uncertainties result from our incomplete understanding of the climate system and our need to incorporate arbitrary physical simplifications to increase the validity of the model's simulations.

Most chapters in this book refer to model results, which are often presented as colored maps that are graphical representations of the output produced by these models. When comparing results from various models, one might think that the models' results diverge. However, in most cases, the results are more similar than different.

Climate models can be tested and 'tuned' with past data (see Chapter 4) to ensure that they can reproduce what we know has happened. These models have many limitations – some are overcome by the continuing increase in computing power – while those related to the limits of predictability of the climate system will remain. At this point, models are the only tool scientists have in attempting to predict future climate. In the future, higher-speed computers might help to resolve the climate forecasting problem.

When running climate models for prediction, scientists must make assumptions about the evolution of human societies both in terms of size and in terms of energy usage and associated emissions. Hence, uncertainties associated with model projections of future climate are due to the imperfection of models and almost equally to uncertainties in human beings' future behavior. (See spread in projected temperatures for 2100 in Figure 1.2.)

The main scientific points

Strong observational evidence of climate change over the past century is presented in this book, including global surface temperature increases, ocean heat content increases, melting glaciers, and rising mean sea level directly associated with an increase in greenhouse gas (CO_2, CH_4, N_2O, O_3 in the troposphere) concentrations and land-use changes (deforestation, desertification, urbanization). It is shown that these recent changes are occurring rapidly over a background of slower changes in both temperature and greenhouse gas concentrations taking place over long timescales, such as between glacial and interglacial periods separated by tens of thousands of years. The additional greenhouse effect is described and demonstrated to be largely responsible for the significant temperature changes observed while also acting in conjunction with feedbacks (e.g. water vapor, cloud, snow/ice, vegetation) in the climate system that either enhance or reduce the changes, thus complicating our understanding of climate change and climate prediction.

Climate models all agree in their projections of increased warming in the future as a result of accumulated heat in the oceans (inertia of the system) and unavoidable greenhouse gas emissions from high levels of fossil energy consumption in both developed and newly developing nations. The tiny particles that make up aerosols have the ability to offset some of the greenhouse-induced warming and potentially mask increased warming that could occur if these aerosol emissions (e.g. pollution) are reduced in the future. Aerosols create an important link between human activity and natural processes through their effects on clouds.

We also learn from this book that the water cycle is a major component of the climate system because it couples the atmospheric branch of climate with other reactive components of climate systems (oceans, terrestrial hydrology, ice sheets, and vegetation). Despite some robust aspects of our knowledge, based for instance on thermodynamic principles, it is and will remain difficult to predict changes to the hydrological cycle on a warmer Earth. Hence, the limited understanding of the hydrological cycle in warmer conditions constitutes one of the major roadblocks to quantitative predictions of climate change.

For the land component of the hydrological cycle, we are presented with results about which there is some agreement among scientists. In regional predictions, for example, it is agreed that likely changes will include more precipitation but less snow over the Arctic, wetting of the tropics, drying of mid-latitudes in some places, and increased drought frequencies in regions already afflicted by pervasive droughts, such as the Sahel in Africa, the Mediterranean region of Europe, and the Southwestern United States.

We learn that the ocean is a dynamic environment, very different from the simple flow depictions of the past. The ocean is a richly intricate physical, chemical, and biological system undergoing continual change. As a result, numerical models still have difficulties making accurate forecasts and, in some cases, even distinguishing between natural climate variability and long-term, human-induced changes.

A look at the carbon cycle shows that the vegetation and the ocean are continuously sequestering, in about similar proportions, a significant part of the CO_2 humans are emitting (with the subsequent acidification of the ocean). Both thus provide incredible assistance in reducing atmospheric CO_2. From the chapter on the carbon cycle, we learn that Earth has not experienced the present atmospheric CO_2 concentration of over 375 ppm for at least 20 million years. This lack of past experience makes it difficult to predict what may happen in the future.

Overall, we see a need for adaptation to both the warming that has already occurred and that which will inevitably continue to occur. We will have to envision mitigation strategies to prevent further large-scale and potentially disastrous changes.

The chapter on the cryosphere, which encompasses the coldest regions of the world, tells us that this is where the most rapid warming will occur: mountain glaciers are retreating, snow cover steadily diminishes from year to year, the Greenland and Antarctic ice sheets show signs of melting, and the ice shelves are collapsing. Less Arctic sea ice and early snowmelt are, via positive feedbacks, augmenting and accelerating global warming rates and sea-level rise. Significant impacts are already observed on polar ecosystems, and highly populated coastal areas are at risk.

As one of the epigraphs at the beginning of this introduction mentions, the Arctic is the 'canary' of climate change. This we believe, and we have consequently devoted an entire chapter to a discussion of the Arctic. This region is experiencing a record loss of sea ice year after year, and scientists worldwide are attentively monitoring these changes. The anticipated effects are well-known: As the Arctic region absorbs more heat from the sun, causing the ice to melt further, the relentless cycle of melting and heating will shrink the massive Greenland glaciers and dramatically raise sea levels. The question that can now be raised in light of the Arctic changes is whether this new acceleration of melting ice, while considering the inertia of the ice sheet, represents a critical and irreversible tipping point beyond which the climate cannot recover.

Facing climate change together

Scientists who have contributed to this book do not intend their contribution to be policy prescriptive but rather policy relevant by providing citizens and policy makers with the best information available for assessing strategies for mitigation of climate change and for adaptation for surviving it. At the same time, these scientists are also concerned citizens who cannot help but express their views on the urgency of this crisis, even if the scientific evidence is not all in. We must limit emissions while adopting strategies to adapt and mitigate the effects of climate change already under way. This agenda will require further development of research activities needed to consolidate basic knowledge, sustain global observing systems that serve both scientists and society, and improve understanding and predicting capabilities through modeling and data assimilation. Scientists can also encourage other citizens to reevaluate their lifestyles, investigate lower carbon-based energy sources and energy-consuming actions, and learn from other countries' efforts to help reduce climate change by initiating innovative societal development.

Europe speaks with a strong voice on the climate change issue. Unfortunately, action is frequently lagging behind the forceful words. France, for instance, can offer only a few examples of innovative action to combat climate change. Under pressure from the automobile industry, the French government recently backed down on its intent to reduce maximum speed on the ever-lengthening freeway network and to impose additional taxes on thirsty four-wheel-drive vehicles. Most public funding in the transportation area goes to highway construction rather than to supporting rail or river freight. France's relatively low greenhouse gas emissions are mostly due to the predominance of nuclear electricity. Little effort has been made toward developing renewable energy sources, apart from biofuels, and R&D funding on low-carbon energy sources is at a critically low level. The stability in emissions level (in conformance with France's commitment to the Kyoto Protocol) results from the balance of decreasing industry and agriculture emissions and the steady increase of transportation and habitat sector emissions. Other countries in Southern Europe are far from filling their Kyoto obligations (Italy: +12% compared to 1990, instead of an allotted −6%; Spain: +48% in 2004 compared to 1990, instead of an allowed +15%), jeopardizing the overall EU reduction goal. Reductions of future emissions by a 'factor of 4' by 2050 are being studied in France and the United Kingdom, helping to provide evidence that this goal is achievable. However, there is a lack of both will and consistency in policies that makes it sound just like a political slogan.

On the other side of the Atlantic, the US government has not ratified the Kyoto Protocol, and US CO_2 emissions continue to grow rapidly: +16% in 2003 compared to 1990, and a projected +54% in 2030. Yet, the United States has a stronger R&D program related to climate and renewable resources. Many states, cities, and local communities in the United States have adopted exemplary strategies to reduce fossil fuel consumption, foster the use of renewable energy, and develop zero-emission homes. The Western Governors' Association unanimously passed a resolution calling on states and cities to reduce human-caused greenhouse gases, a significant step for governors of states with power plants or major coal, oil, and gas reserves. In the eastern part of the North American continent, the six New England states of the US and the four maritime provinces of Canada have committed to participate in devising regional goals and work with other states and provinces in the region to meet these goals. In California, the governor promulgated a law in 2006 enforcing greenhouse gas emission limitations in this state, responsible for 9% of the global CO_2 emissions.

In a world in which the past 40 years have seen a widening gap between rich and poor, the prospects of climate change are especially daunting. Poor countries, and even some regions of more developed countries, could find themselves overwhelmed with massive numbers of 'climate migrants' fleeing regions repeatedly struck by extreme events. Yet, climate disasters may hit both poor and rich countries. A preview of what may become a common occurrence in the future was seen when Hurricane Katrina devastated New Orleans, Louisiana, in 2005. Poor communities were most affected by the disaster. Many people died, and large numbers of survivors were still looking for a place to settle almost one year later. Another example is the heat wave that struck Western Europe in August 2003, with more than 15 000 deaths, mostly among elderly and poor people, in France alone. The world needs to better prepare itself for the inevitable displacement of millions of people in the future. In fact, because of rapid population growth and coastal development, the world is not even prepared to deal with 'normal' levels of climate variability.

The contributing authors hope that this book will help convey the need to combine efforts in improving our capacity to understand and predict climate change while working to maintain a safe and healthy environment for future generations. The United States and France (and more generally Europe) have the opportunity to play a leading role on the path toward sustainability and environmental responsibility. Although decisions about how to minimize and adapt to climate change will ultimately be based on social, economic, and political policies, it is absolutely vital that any analysis of climate change be based on sound science.

More and more people are tempted to turn to technologically based 'solutions,' recognizing the difficulties of changing behaviors. These are collectively named 'climate engineering.' They include iron fertilization of the oceans, injection of sulfate aerosols in the atmosphere to reduce incoming solar radiation, or even deployment of large mirrors in orbit to deflect part of the solar flux. Although developing research on these topics is legitimate, it may open a Pandora's box. By no means should research on climate engineering, in our opinion, be used as an excuse to discontinue or even delay the immediate efforts needed to implement 'carbon-light' solutions. The involuntary experimentation under way on climate has proven risky enough to refrain from initiating new, adventurous alterations on a complex climate system not fully understood.

Our global society is now confronted with the urgent need for a wise response to climate change. Unlike some ancient civilizations that presumably collapsed in part due to excessive pressure on their environment, our society benefits from an early warning and predictive tools to help society adjust its behavior and prevent a disastrous outcome. The question now is: Will we be able to do better?

Recommended books and articles

J.-L. Fellous, *Avis de Tempêtes – La Nouvelle Donne Climatique* (Paris: Odile Jacob Sciences, 2003).

S. R. Weart, *The Discovery of Global Warming* (Cambridge: Harvard University Press, 2003).

E. Kolbert, *Field Notes from a Catastrophe: Man, Nature, and Climate Change* (New York: Bloomsbury Publishing, 2006).

1

The global consensus and the Intergovernmental Panel on Climate Change

RICHARD C. J. SOMERVILLE AND JEAN JOUZEL

After all, what's the use of having developed a science well enough to make predictions, if in the end all we're willing to do is stand around and wait for them to come true?
—F. Sherwood Roland, 1995 Nobel Laureate in Chemistry

C'est ainsi que la température est augmentée par l'interposition de l'atmosphère, parce que la chaleur trouve moins d'obstacle pour pénétrer l'air étant à l'état de lumière, qu'elle n'en trouve pour repasser dans l'air lorsqu'elle est convertie en chaleur obscure. [It is such that the temperature is increased by the interposition of the atmosphere, because the heat faces less obstacle to penetrate the air when it is in the form of light than it finds to go through the air, when it is converted to dark heat.]
—Joseph Fourier, *Mémoire sur les températures du globe terrestre et des espaces planétaires*, Mémoires de l'Académie des sciences

Climate change science, a summary of what we have learned

Climate is complex, and the scientific study of climate change begins with observations of the climate system. Scientists have made dramatic progress

---

Richard C. J. Somerville, a climate theorist, is Distinguished Professor Emeritus at Scripps Institution of Oceanography, University of California, San Diego. He was a coordinating lead author for the 2007 Fourth Assessment Report of the Intergovernmental Panel on Climate Change.

Jean Jouzel, a geochemist with the French Atomic Energy Agency (CEA), is involved in the study of ice cores from Antarctica and Greenland to reconstruct past climate changes. He was lead author of the second and third Intergovernmental Panel on Climate Change (IPCC) reports and is currently member of the IPCC Bureau and vice-chair of its scientific group (group I). Since 2001, he has been Director of Institut Pierre Simon Laplace, the leading French institute in the study of global environment.

in making many kinds of climate observations and analyzing them. As a result, we now possess a more certain and clearer understanding of the climate change that has already occurred, especially in recent decades. For example, we know that the world has warmed during the twentieth century by about 0.6 °C (1.1 °F). This number is a global and annual average of atmospheric temperature near the Earth's surface over land, combined with the temperature of the ocean surface. Obtaining this single number for the average temperature of our planet involves carefully analyzing a complicated mix of different kinds of data. Thermometers at weather stations are used together with data from ships, satellites, and other sources. Several independent groups of researchers have performed such analyses. The result is a well-calibrated number for the Earth's present-day average temperature and a trustworthy record of how this fundamental parameter has changed over time.

We know too that, during the period in which instrumental records have been adequate to characterize this global average temperature – extending from about 1861 to the present – the 1990s was the warmest decade on record, and that 1998, which was an El Niño year, has been the warmest year, at least in the twenty-first century. Recently, we learned that 2005 was about as warm as 1998.

For the period before about 1861, instrumental records are inadequate to determine a global average temperature, and other climate parameters are even less well known; thus our knowledge of climate change in this preinstrumental period relies on so-called proxy data, such as tree-ring records. These data allow us to extend our knowledge of the Earth's climate history deep into the distant past. However, proxy data have their own uncertainties and must be interpreted carefully. In addition, only limited geographical areas are accessible with specific proxy data. Nevertheless, we can say that the Northern Hemisphere is likely to have been warmer during the twentieth century than during any other century of the last 1000 years. Less is known about the Southern Hemisphere, which is less observed because it has much more ocean and less land area than the Northern Hemisphere.

In the latter half of the twentieth century, instrumental coverage became adequate to suggest that, on the average, daily minimum air temperature values, which normally occur at night, were increasing about twice as rapidly as daily maximum temperatures. However, the average diurnal range of temperatures has not changed from 1979 to 2004. During the same period, sea surface temperatures increased about half as much as near-surface air temperatures over land. These changes in sea surface temperatures are consistent with expectations for a planet gradually warming because of a strengthened greenhouse effect.

Many other aspects of observational evidence for climate change have also been studied. For example, the geographic extent of snow and ice cover has decreased globally in recent decades. During this period, the heat content of the ocean has also increased significantly, and sea level has risen. The twentieth-century rise in sea level has been estimated at between 0.1 and 0.2 m, on global average.

Partial or less-certain information has been obtained regarding many other features of climate. Over some Northern Hemisphere land areas, precipitation has increased. However, limited instrumental coverage over the oceans and the Southern Hemisphere makes assessing precipitation changes there more problematic. Since the mid 1970s, the complex climate phenomenon known as the El Niño–Southern Oscillation, or ENSO, has undergone changes. In recent decades, ENSO episodes have occurred more frequently than in the preceding decades. Also, they have tended to last longer and have been more intense.

Forcing agents responsible for climate change

In parallel with observational studies documenting the reality of climate change, other research has increased our knowledge of the causes of this change. Scientists speak of 'forcings' to denote the external factors that can give rise to climate change. These include natural forcings, such as volcanism and changes in the Sun, as well as human-caused (or anthropogenic) forcings, such as adding to the amounts of atmospheric gases that contribute to the greenhouse effect. In addition, climate change can occur without forcings, simply because of the internal variability of the climate system itself.

An especially well-documented forcing is that resulting from changes in the amount of carbon dioxide in the atmosphere. During the last 250 years, this amount has increased by more than 30%. We know this because we have seen a combination of highly accurate direct measurements of carbon dioxide, beginning in the 1950s, together with the analysis of older samples of air trapped in ice in Antarctica and Greenland. Thus, it has been possible to connect the modern instrumental record with a long series of determinations of carbon dioxide concentrations extending back many centuries. As a result, we can say confidently that the atmospheric concentration of carbon dioxide now is higher than at any other time in the past 420 000 years. In fact, during the last 20 million years, the Earth has probably never experienced higher atmospheric carbon dioxide amounts than at present.

In a very real sense, the modern concern about anthropogenic climate change can be traced directly to these carbon dioxide measurements, and in particular to the instrumental record established by the late Charles David Keeling

(1928–2005). It was Keeling who designed and built the first highly accurate instrument for measuring atmospheric carbon dioxide in the 1950s and then established the increasing trend in concentration. Isotopic evidence demonstrates its anthropogenic origin, because carbon dioxide from fossil fuels has a clear isotopic 'signature.'

Why are we so confident that humans are the cause of the increase in carbon dioxide? To give a brief description of part of the answer, it is progress in fundamental chemistry that has allowed researchers to meet this particular challenge. The isotopic analysis of atmospheric air is a fascinating example of how research progresses, and in particular, how scientists have been conclusively able to accomplish seemingly difficult feats, such as attributing the increase of atmospheric carbon dioxide to human activities. Almost all chemical elements occur in the form of more than one isotope, that is, as atoms of the same element that have different masses. Carbon occurs in the form of isotopes with atomic masses 12, 13, and 14. The isotope with mass 12 accounts for nearly all of the carbon found in nature, but the form with mass 13 is also found in atmospheric carbon dioxide.

The carbon dioxide produced from burning fossil fuels or forests has a different isotopic composition than that from other sources of carbon dioxide in the atmosphere. This is because growing plants, which extract carbon dioxide from the atmosphere through photosynthesis, have a preference for the lighter isotope (the one with mass 12). Because fossil fuels originate in plants from the distant past, when we burn these fuels, they produce carbon dioxide with more of this isotope. By analyzing the amounts of the different isotopes in air samples, scientists have confirmed the fossil fuel origin of most of the added carbon dioxide.

For example, we now know that during about the last 20 years of the twentieth century, about three-quarters of the anthropogenic emissions of carbon dioxide into the atmosphere can be attributed to fossil fuel burning. Most of the rest resulted from land-use changes, predominantly deforestation.

Carbon dioxide contributes to the greenhouse effect, but so do several other gases. During the last 250 years, the amounts of these other 'greenhouse gases' have also been increasing. Methane increased by about 150% during this period. About half of the increase is due to human activities, including fossil fuels, cattle, rice agriculture, and landfills. Nitrous oxide increased by about 17% during this period, partially because of human activities. A number of other greenhouse gases, including the chlorofluorocarbons and related chemical compounds, are due entirely to human activities. These gases do not exist in nature but have been manufactured and widely used in applications ranging from refrigeration to the manufacture of electronic components.

Until recently, about half of the contribution of human activities to increasing the greenhouse effect could be attributed to carbon dioxide, and the combined effect of all the other anthropogenic emissions of greenhouse gases made up the other half. Much more accurate determinations of these contributions can be found in the scientific research literature and has been assessed in the IPCC reports.

In addition to gases, microscopic liquid and solid particles in the atmosphere, called aerosols, also can have a substantial effect on climate change. Anthropogenic aerosols are short-lived, typically remaining in the atmosphere for several weeks, compared with greenhouse gases such as carbon dioxide, which have characteristic atmospheric residence times measured in decades to centuries. Furthermore, unlike greenhouse gases, which warm the surface of the Earth by strengthening the greenhouse effect, atmospheric aerosols tend to cool it, either by reflecting sunlight back to space or by absorbing sunlight. Both processes reduce the amount of sunlight reaching the Earth's surface.

Scientists use the concept of 'residence time' to characterize how long aerosols and gases typically remain in the atmosphere before being removed by natural processes. For example, some pollutant particles tend to be quickly washed out of the atmosphere by rainfall, but a carbon dioxide molecule might remain in the atmosphere for decades or centuries. Because of their relatively short atmospheric residence times, aerosols, unlike the greenhouse gases, have concentrations that are highly variable in space and time, with the largest concentrations occurring near the sources of the aerosols. Many aerosols are also pollutants that affect air quality, and some are responsible for acid precipitation and related harmful effects. The major sources of anthropogenic atmospheric aerosols are fossil-fuel use and biomass burning.

In addition to their direct effect on solar radiation, aerosols may also have indirect effects operating through clouds. For example, aerosols serve to nucleate cloud particles, thereby affecting properties of cloud particles and the ability of the cloud to interact with radiant energy from the Sun and the Earth, as well as precipitation processes and the lifetime of clouds. Our understanding of aerosol effects on climate, especially these indirect effects, is generally much less certain and complete than our understanding of the effects of greenhouse gases.

In addition to the anthropogenic effects of greenhouse gases and aerosols and other human activities such as land-use changes, climate over the last century has also responded to so-called 'natural' forcings. 'Natural' in this context refers to causes other than human activities. The two natural forcings that are thought to be most important on this time scale are solar variability and the effects of aerosols originating from explosive volcanic activity. Compared with anthropogenic (human-caused) forcings, these natural forcings are thought to have been responsible for a smaller effect on climate. During the last two decades of the twentieth century, and possibly the last four decades, the result of these natural forcings was a net cooling effect.

During the first half of the twentieth century, changes in the amount of energy emitted by the Sun are thought to have been responsible for warming the climate. This finding is uncertain, however, because accurate measurements of solar irradiance were not yet possible at that time. Satellite measurements of solar irradiance began in the late 1970s and have shown small periodic variations because of the 11-year solar cycle. Major volcanic eruptions took place in the period from 1880–1920 and also in the period from 1960–1991. The cooling due to an explosive volcanic eruption which gives rise to large amounts of aerosols in the stratosphere may be substantial and may last for several years, as was the case for the Mount Pinatubo eruption of 1991.

Computer simulation of the climate system and future climate change

Science strives to be quantitative. For example, it is relatively easy to predict that, far from equatorial regions, summer will definitely be warmer than winter, but exactly how much warmer will it be? In a specific place, will summer be warmer on average by $0.5\,°C$, or 2, or 7, or $20\,°C$? In the same way, a basic understanding of the greenhouse effect allows us to be quite confident that adding carbon dioxide to the atmosphere will lead to a warmer world. The much more important and difficult question is how much warming can be expected for a given increase in carbon dioxide. In addition, climate change involves much more than a simple warming. The warming may be thought of as just a characteristic of climate change in the same way that a fever is a symptom of disease, but it is only a partial indication that does not tell the full story. In the case of a warming world, we also want to know what changes to expect in rainfall, sea level, storminess, and many other important aspects of climate.

The aspect of climate science that is concerned with the range of possibilities for climate change in coming decades relies heavily on results from computer simulations of the climate system. The key tool in this type of research is the global climate model, or general circulation model (GCM). This class of models has been developed to incorporate the most important features of the climate system. The climate system is made up of the global atmosphere, oceans, land surfaces, ice and snow, and many of the interactions between these components,

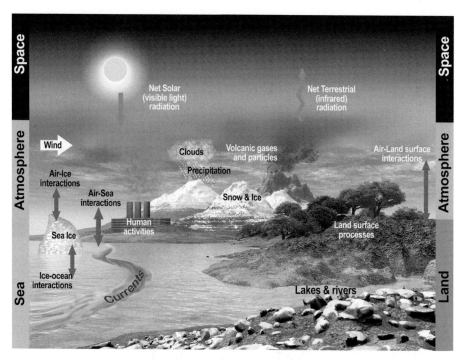

Figure 1.1 A simplified rendition of the climate system components and their interactions. (For image in color, please see Plate 1.)

including some biological and chemical processes. See Figure 1.1. Such models require interdisciplinary teams of scientists to create and use the models, as well as substantial computer resources to run them.

Computer simulation plays a central role in many areas of science, and it is essential in climate science because of the impossibility of performing large-scale controlled experiments on the entire Earth system. In this way, climate science differs from traditional laboratory sciences and resembles areas of science such as astronomy. As a substitute for direct experimentation, climate models allow researchers to explore the consequences of hypothetical future scenarios involving various forcings, such as prescribed emissions of greenhouse gases in the future. The GCM in practice is a large computer program incorporating much of our current understanding of the climate system, expressed as equations representing relationships between variables such as temperature, winds, ocean currents, and greenhouse gases. This program can be altered to permit the investigation of how the modeled climate would respond to such forcings scenarios.

Ideally, one would want a GCM to be based entirely on well-understood physical laws. In practice, however, the complexity of the climate system and our incomplete understanding of it mean that GCMs also incorporate somewhat arbitrary physical simplifications to increase the realism of their simulations of the present observed climate. Furthermore, the challenge of finding solutions to complicated mathematical equations using computers also involves approximations and introduces some inevitable errors. It is also true that our imperfect observational knowledge of climate limits our ability to simulate it with numerical models. Thus, GCMs are useful tools, but not perfect ones.

Modern climate research makes use of a hierarchy of models covering a wide range of complexity. Some models are highly idealized and simulate a small number of specific physical processes in detail. GCMs are the most comprehensive and ambitious models in the hierarchy. These models are continuously under development and are revised as improved knowledge of specific aspects of climate becomes available. The most realistic modern GCMs are effectively a synthesis of much of our present understanding of the climate system.

Although climate science, like all science, is incomplete in some respects and is subject to various uncertainties, many lines of evidence support the conclusion that models have now reached the stage of development at which their ability to simulate many aspects of the climate system makes us increasingly confident in at least the broad outlines of model projections of future climate change.

Recently, climate modeling groups have attempted to simulate the evolution of climate over the twentieth century with several GCMs, using various combinations of natural and anthropogenic forcings as inputs to the models. A clear conclusion of this research is that models driven with only natural forcings (solar variability and volcanic aerosols) do not accurately simulate the observed changes in climate over the twentieth century, nor do models driven with only anthropogenic forcings (human-caused greenhouse gases and aerosols). However, when the natural and the anthropogenic forcings are used together as inputs to the models, the most realistic twentieth-century climate simulations are obtained.

Models can also be used to explore and assess the sensitivity of the climate system to specific physical and biogeochemical processes. As a result of many studies of this type, climate researchers have identified particular aspects of the climate system that are responsible for much of the remaining uncertainty in projecting future climate change. Among the key processes involved in determining the sensitivity of the climate system to greenhouse gases are several complex and incompletely understood processes.

These aspects of climate models, which are high-priority candidates for future research, include interactions and feedbacks between clouds and radiative energy transfer (involving both sunlight and the heat energy emitted by the Earth). Many climate scientists consider these cloud-radiation processes, which in principle could either amplify or reduce the effects of increasing greenhouse gas amounts, to be the single most important source of uncertainty in forecasting future climate change on time scales of decades to centuries. Other critical processes involve air–sea interactions, the dynamics of ocean circulation, the stability of the large ice sheets on Greenland and Antarctica, and many aspects of the role of aerosols in the climate system. To reduce the uncertainties and intermodel differences in climate simulations and projections, better observations are necessary, and further research on these topics is urgently needed.

Human influence on past and future climate change

The Intergovernmental Panel on Climate Change (IPCC), which is discussed in detail later in this chapter, has issued four major Assessment Reports, in 1990, 1995, 2001, and 2007. This series of reports has been characterized by increasingly strong statements regarding the role of human activities as a causal factor in climate change. The reason for this increase over time in the certainty expressed by the IPCC assessments is a result of both scientific progress and of the changing climate itself. As climate change science has advanced with the passage of time since the formation of the IPCC in 1988, and as humanity has added more greenhouse gases and aerosols to the atmosphere and otherwise modified the planetary environment, the evidence has become essentially undeniable that human activities are now responsible for many of the observed changes in climate.

In the language of climate change science, this topic is known as 'detection' and 'attribution.' Here, detection is used in the specific sense of determining whether a particular observed aspect of climate, such as global mean surface temperature or sea ice extent, has changed in a well-defined statistical sense. Detection is concerned with whether such a change is likely to provide a statistically significant signal that can be distinguished from the inevitable noise of natural climate variability. Detection is not concerned, however, with providing a reason for such a change. Attribution then refers to the task of establishing the causes of such a detected change, with a defined level of confidence.

Thus, it is especially noteworthy that the IPCC, in its First Assessment Report (FAR, [1]) in 1990 did not make any specific assertion about whether the anthropogenic signal, or 'fingerprint,' could yet be distinguished from the background noise of natural climate variability. The IPCC Second Assessment Report (SAR, [2]) in 1995, however, did state the carefully worded conclusion, 'The balance of evidence suggests a discernible human influence on global climate.'

In 2001, however, the IPCC Third Assessment Report (TAR, [3], [4]) reached the following unequivocal conclusion: 'There is new and stronger evidence that most of the warming observed over the last 50 years is attributable to human activities.' Among the links in the chain of evidence leading to this result, the IPCC cited in the TAR a longer and more carefully examined temperature record, better estimates of natural variability in the climate system, and progress in the scientific investigation of detection and attribution.

The IPCC, which assesses research but does not itself perform research, avoids making specific attempts to forecast future climate change. One reason for not trying to predict climate is the great difficulty in predicting factors such as population, economic development, and other circumstances that will strongly affect emissions of greenhouse gases, aerosols, and other anthropogenic inputs to climate change. Instead, the IPCC has adopted the practice of developing scenarios depicting a range of possible emissions outlooks. These scenarios are laid out in an IPCC Special Report on Emissions Scenarios (SRES). Major climate modeling groups then use these SRES scenarios as inputs to their GCMs. Thus, the model results are not forecasts of either human behavior or of climate change. Instead, they are explorations of hypothetical futures and are in the nature of responses to questions such as, 'If the following specific emissions scenario were to occur in reality, how would the climate system respond to it?'

The TAR concluded, however, that it was virtually certain that human activities throughout the twenty-first century would continue to change the chemical composition of the global atmosphere. In particular, IPCC noted in the TAR that atmospheric carbon dioxide concentrations were virtually certain to continue to increase with time during the twenty-first century because of continued fossil fuel burning. Over the range of SRES scenarios, the TAR found that, in 2100, atmospheric carbon dioxide concentrations might be expected to range from 540 to 970 parts per million by volume (ppmv). Such concentrations represent an increase from 90% to 250% from the baseline preindustrial value of 280 ppmv.

By 2100, global average surface temperatures, according to the TAR, might be expected to increase from 1.4 to 5.8 °C, relative to 1990 values. This range includes the climatic response to the entire range of SRES scenarios as determined by several climate models. The TAR also discusses the variety of likely consequences for other climate variables, including sea level, precipitation, and El Niño occurrence, and the uncertainties associated with these outlooks. Sea level, for example, might be expected to rise from 0.09 to 0.88 m above 1990 values. This large range reflects the breadth of the emissions estimates included in the SRES scenarios and the different responses of different GCMs to forcings, as well as the scientific uncertainties associated with the dynamics of Greenland and Antarctic ice sheets and other factors affecting sea level rise.

The Summary for Policymakers based on the Working Group (WG) I (physical climate science) contribution to the latest IPCC assessment report released by the IPCC in early 2007, and later in the full report, concludes that recent warming of the climate system is 'unequivocal' and that most of it is at least 90% certain to be the result of human activities. WGs II and III, dealing with mitigation and adaptation and related issues of climate change, released their contributions to the IPCC Fourth Assessment Report later in 2007.

Among the many scientific findings highlighted in the Summary for Policymakers of the WG I contribution to the new IPCC report are the following:

A. Some observational evidence of recent climate change

1. The largest CO_2 growth rate in the instrumental record is found in the most recent decade.
2. The Earth is now globally about 0.76 °C warmer than in the late nineteenth century.
3. The linear temperature trend in the last 50 years is nearly twice that for the last century.
4. North Atlantic hurricanes have intensified since 1970.
5. Arctic temperatures have increased at about twice the global rate.
6. Arctic sea ice has shrunk by about 2.7% per decade.
7. Globally, 11 of the last 12 years are among the 12 warmest on record since 1850.
8. The global ocean is warming to depths of at least 3000 m.
9. The ocean has absorbed more than 80% of the heat added to the climate system.
10. Sea level rise averaged globally over the twentieth century has been about 17 cm.

B. Some projections of future climate change

1. Sea level will rise about 0.2 to 0.6 m in the twenty-first century (with caveats: we cannot yet assess the potential for further sea level rise caused by ice sheet dynamics because of a lack of sufficient scientific understanding).
2. Larger values of sea level rise cannot be excluded (125 000 years ago, sea level was 4 to 6 m higher than at present, but high temperatures then were sustained for centuries).
3. During the next 2 decades, we expect about 0.2 °C per decade of further warming. This anticipated future warming continues observed recent trends, which are themselves consistent with earlier IPCC projections.
4. Ocean acidity will increase, with a further reduction of 0.14 to 0.35 pH units by 2100.
5. Snow cover and sea ice will continue to contract.
6. Heat waves and heavy precipitation will become more frequent.
7. Tropical cyclones will become more intense.
8. Warming and sea level rise will continue for centuries.

9. The Atlantic meridional overturning circulation will slow.
10. Precipitation will tend to increase in high latitudes.
11. Precipitation will tend to decrease in subtropical latitudes.

The Intergovernmental Panel on Climate Change

The IPCC and the interface between science and public policy

The results of the modern science of climate change, as summarized in this and the other chapters of this book, are impressive but also sobering. Taken together, these results are persuasive in that they include, first, a large body of observational evidence characterizing the climate change that has already occurred. In addition, the science has developed to the point at which we are often able to connect the observed changes with the causes that are responsible for them. The resulting picture demonstrates that human activities dominate in causing recently observed climate change. Finally, our simulations of plausible futures indicate the strong possibility of severe climate change in coming decades if human activities continue on their present path.

These projections, obtained from increasingly sophisticated climate models are, in a certain sense, definitive. We clearly are at the dawn of a century which, if we do not take care, may experience a very rapid warming with consequences that are still poorly defined, but this book lets us glimpse that they risk being very harmful.

In fact, climate scientists are not the first to address a problem of planetary scope originating in human activities. During the 1970s, atmospheric chemists were concerned about the potential role of supersonic aircraft on the stratospheric ozone layer, the preservation of which is essential for our health and, more generally, for life on Earth. Chemicals called chlorofluorocarbons (CFCs) rapidly proved to be the most worrisome, so much so that the United Nations Environment Programme (UNEP) established a Coordinating Committee on the Ozone Layer in 1977. Then, in 1985, an international conference in Vienna resulted in a convention for the protection of the ozone layer. This decision was facilitated by the publication that same year of the discovery of the hole in the ozone layer above Antarctica. This discovery resulted from observations made at the British Antarctic station at Halley Bay by a team led by Joseph Farman. Additional scientific proof accumulated and led 24 countries to sign, on September 16, 1987, the Montreal Protocol on Substances that Deplete the Ozone Layer. This protocol anticipated measures to control and subsequently eliminate certain chemical compounds. Because of successive amendments, the agreement was gradually extended to cover the entire group of substances that threaten the ozone layer. This battle is now well on the way to being won, with the hope of a complete recovery of the ozone layer by approximately the middle of the twenty-first century.

The threat posed by global warming is of a completely different character, because in the case of the ozone layer, several major corporations quickly developed ozone-safe substitutes for CFCs, while, on the other hand, the increase in the greenhouse effect appears essentially linked to the development and prosperity of human society. Even so, the Montreal Protocol undeniably played the role of precursor to the agreement that would emerge a decade later at Kyoto.

The debate about the strengthening of the greenhouse effect linked to human activities actually had its origins much earlier than the debate concerning the ozone layer. A century would pass between the first estimate of the consequences of increasing the concentration of carbon dioxide in the atmosphere, published by the Swedish scientist Svante Arrhenius (1896), and the negotiation of the Kyoto Protocol. However, it was not until relatively recently that the scientific community would understand the magnitude of the potential consequences of human activities on our climate. This realization required waiting until the 1970s, when the availability of supercomputers made possible the first model simulations of climate. These calculations suggested that a significant warming, between 1.5 and 4.5 °C (2.7 and 8 °F) would result from doubling the atmospheric CO_2 content.

The first World Climate Conference, held in Geneva in 1979, and then another conference, in Villach, Austria, in 1985, shed light on the risk of a climate warming associated with an increase in the greenhouse effect. These conferences led to the creation, in 1988, of the IPCC, under the auspices of UNEP and of the World Meteorological Organization (WMO). The objective of the IPCC was to carry out, based on available scientific information, an assessment of the scientific, technical, and socioeconomic aspects of the climate changes that might result from human activities. From the very beginning, the IPCC was organized into three working groups (WGs). In this chapter, we confine our interest to WG I, which addresses the scientific aspects of climate change. The other two WGs are concerned with the consequences, adaptation, and vulnerability aspects of climate change (WG II) and with mitigation aspects of climate change (WG III).

The IPCC has become the de facto voice of the mainstream scientific community, as the world seeks to understand the findings of climate science and their relevance to public policy. The IPCC has been immensely influential in the debate about anthropogenic climate change and has had a significant effect in motivating efforts to limit the emissions of gases that increase the greenhouse effect and cause global warming. The IPCC was established to provide an authoritative assessment of results from climate science as input to policy makers. Its mandate is to assess research, not to do research. Its reports are policy-relevant but not policy-prescriptive. Thousands of scientists throughout the world have contributed to the IPCC effort.

The four IPCC Assessment Reports (1990, 1995, 2001, 2007)

In 1989, the United Nations General Assembly asked the IPCC to present its first report the following year. Published in 1990, this First Assessment Report (FAR, [1]) gave an account of what was known regarding the role of human activities in modifying the composition of the atmosphere through carbon dioxide, methane, and other gases contributing to the greenhouse effect. The FAR also described extremely worrisome predictions for the twenty-first century, with an average global warming that might reach 3 °C (5.4 °F) by 2100, accompanied by a sea level rise of 65 cm. Although this report described the numerous uncertainties accompanying these climate projections based on computer models, and although it was careful to avoid attributing the warming of 0.3 to 0.6 °C (0.5–1.1 °F) that had been observed over the preceding 100 years to a strengthened greenhouse effect, the FAR would nevertheless come to play an extremely significant role in the debate on climate change.

In fact, based largely on the assessment described in the FAR, the United Nations became involved in the negotiation of a Framework Convention on Climate Change (UNFCCC), which was completed at the Earth Summit in Rio in 1992. Today, some 189 countries are signatories or 'parties' to the UNFCCC, which has as its objective: 'To stabilize the concentrations of greenhouse gases in the atmosphere at a level that will prevent dangerous anthropogenic interference with the climate system.'

The role of the IPCC is to provide the parties to the UNFCCC, which are governments, with the necessary information concerning the climate-change problem. It is the task of these parties to negotiate and eventually make the necessary policy decisions. This process occurs on an annual cycle, governed by a series of annual meetings, each called a Conference of the Parties (COP). This process will be familiar to all those who are interested in the evolution of our climate. This synergy between the IPCC and the UNFCCC, bearing the fingerprint of mutual interdependence, seems to function in an entirely exemplary fashion.

The publication in 1995 of the IPCC Second Assessment Report (SAR, [2]), accompanied by a Synthesis Report summarizing the conclusions of WG I, II, and III, is one example. It reinforced the conclusions of the FAR with respect to the role of human activities in strengthening the greenhouse effect, as well as the observed climatic warming, the projected future changes, and the existence of uncertainties. It called attention to the role of aerosols of anthropogenic origin, which partially compensate for the greenhouse effect. However, the most outstanding conclusion of the SAR was that 'the balance of evidence suggests a discernible

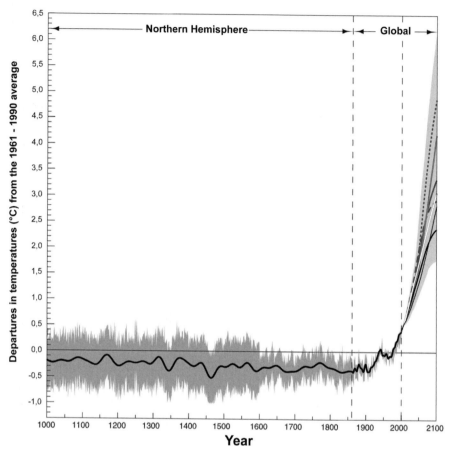

Figure 1.2 Observations since the year 1000 (proxies from 1000 until 1860, instrumental measurements since 1860) over the Northern Hemisphere, and global projections of temperature anomalies until 2100, according to various SRES (Nakicenovic and Swart 2005) scenarios (after IPCC TAR). Intergovernmental Panel on Climate Change (IPCC); SRES, Special Report on Emissions Scenarios; TAR, Third Assessment Report.

human effect on climate.' This guarded and carefully worded conclusion of scientists had a substantial impact at that time and is one of the factors that led in 1997 to the signing of the Kyoto Protocol.

The IPCC Third Assessment Report (SAR, [3], [4]), published in 2001, confirmed this assessment, thanks to improvements in models and the availability of new observational data that provided improved knowledge of climate variability in recent centuries. Thus, the combined efforts of paleoclimatologists, who have reconstructed several climatic time series using complementary methods, and of statisticians, who have combined these results and obtained average values, have led to the publication of a single curve depicting the time evolution of the climate of the Northern Hemisphere throughout the last millennium. See Figure 1.2. This curve remains subject to significant uncertainty, but it leaves little doubt that the recent warming is a departure from natural variability. Climate models confirm this conclusion, with long runs showing that the warming of the last 100 years is, in all probability, not due uniquely to natural causes. In particular, the warming of the last 50 years cannot be explained without taking into account, as a dominant factor, the anthropogenic strengthening of the greenhouse effect. Thus, the TAR concluded: 'There is new and stronger evidence that most of the observed warming observed over the last 50 years is attributable to human activities.'

The IPCC Fourth Assessment Report, published in 2007, provides, like its predecessor, a definitive source of up-to-date information on climate change science. Its Summary for Policymakers states: 'Warming of the climate system is unequivocal, as is now evident from observations of increases in global average air and ocean temperatures, widespread melting of snow and ice, and rising global average sea level.' It states also: 'Most of the observed increase in globally averaged temperatures since the mid-twentieth century is very likely due to the observed increase in anthropogenic greenhouse gas concentrations.' (Here, 'very likely' is precisely calibrated language denoting expert judgment of a likelihood of 90% or more).

IPCC and the policy debate regarding human effects on climate change

Thus, a substantial amount of scientific uncertainty and hesitation at the time of the SAR in 1995 had become greatly reduced by the time the TAR appeared in 2001. The group of 'greenhouse contrarians' grew smaller, so one consequence of the IPCC's conclusion was to relegate the scientific debate to the background and thus pave the way for policy proposals to deal with the perceived threat of human-caused climate change. In the view of many decision makers, the early questioning and doubt regarding the human effect on climate were replaced by a scientific near-certainty, and this change of attitude played a key role in the creation and ultimately the ratification of the Kyoto Protocol.

This protocol required the developed countries, during the years before the 2008–12 time frame, to have reduced their greenhouse gas emissions by 5.2% with respect to 1990 levels. The developing countries, on the other hand, were not required to take any action. The goals also differed from one country to another. For example, France, which already had a relatively low level of greenhouse gas emissions because of its heavy reliance on nuclear power, was required simply to keep its emissions constant, while the United States accepted in principle a reduction of 7% in its emissions. For the protocol to take effect, however, it was necessary that it be ratified by 55% of the member states of the United Nations, which together were responsible for at least 55% of global total emissions. Despite the refusal of the United States to ratify the protocol, this criterion was reached in November 2004 with the signature of the Russian Federation, motivated at least partially by a discount advantage on the emissions permits market favoring countries with economies in transition. Three months later, on February 16, 2005, the Kyoto Protocol officially entered into force.

The IPCC diagnosis set forth clearly in the TAR, that human activities were already modifying our climate, was not the only concern. In addition, there were also alarming predictions of the average increase in temperature by 2100 because of the inertia of the climate system, which implied that the effects of warming would be felt long after the time when humanity succeeded in stabilizing the greenhouse effect.

These conclusions were examined in great detail by the scientific community, by scholarly organizations, by major corporations, and by the media. Greenhouse skeptics questioned, among other things, the validity of climate reconstructions for the last millennium and an apparent disparity between temperatures observed at the Earth's surface and those measured in the atmosphere. These criticisms received attention, especially from the media, but the conclusions of the IPCC have been generally confirmed and repeatedly endorsed. They were supported in 2001 by the United States National Academy of Sciences in a report requested by the White House. In 2005, 11 national science academies of major countries, including the United States, Russia, China, and India, co-signed a declaration calling on world leaders and in particular the major developed ('G8') countries to 'acknowledge that the threat of climate change is clear and increasing.' Several large corporations (e.g. BP) followed closely behind, and the ranks of the skeptics were gradually reduced as this awareness slowly won over the general public.

The IPCC process

The IPCC reports are thus undeniably authoritative, both because of the quality of the scientists involved in writing them and because of the rigorous editorial process the reports undergo. Each report is divided into chapters for which the initial drafting is entrusted to a team of approximately a dozen researchers from different countries, who request input from scientists involved in the relevant disciplines. In addition to these extremely voluminous reports (nearly 1000 pages for the more recent Assessment Reports of WG I), summaries of about 50 pages are also prepared, along with a 'Summary for Policymakers,' which is much shorter and written in a readily accessible style. The entire report is encapsulated in a 'Synthesis Report.' These various documents are all extensively reviewed by both the scientific community (peer reviewers) and by representatives of government authorities. An iterative cycle of reviews and revisions of successive drafts takes place under strict guidelines stressing transparency and broad participation throughout the process.

According to established IPCC procedures, the WG I portion of the full Fourth Assessment Report was developed in the form of successive drafts. These drafts were subjected to formal and fully documented expert and government review processes, during which many thousands of review comments were responded to. The report assessed the peer-reviewed research literature published prior to mid-2006.

The Summary for Policymakers of WG I was first drafted by a team of IPCC scientist-authors and then refined by government representatives in a collaborative process with the scientists. At all stages, including at the final plenary in Paris, the scientist-authors had substantive control over the text. The scientists who headed the teams writing each chapter of the full WG I report were all present in Paris. At the plenary in Paris, governments did not alter the substance but did help to refine the language, with the intended readership of policy makers in mind, so that complicated issues were explained more clearly and in more accessible terms. The language in the final document was negotiated and then unanimously approved line by line by the 113 governments represented at the plenary.

As a result of this process, the Summary for Policymakers is based firmly on the underlying chapters of the full report, which in turn are rooted in the assessed peer-reviewed scientific literature. The wide participation of the scientific community, the insistence on the highest standards of scientific accuracy and rigor, and the absence of any policy prescription are the characteristics that render the IPCC Assessment Reports so powerful. This is precisely why these reports serve a unique role in informing policy makers, as well as other interested communities such as businesses, media, and the broad public. These reports are widely regarded as authoritative statements summarizing the best available scientific information on climate change and are widely used as textbooks and reference material in educational applications.

Finally, a complete set of technical reports is published at intervals of 5 or 6 years. The IPCC reports are available in print versions (see references), and several recent IPCC reports are also available as free downloads from the IPCC website: www.ipcc.ch.

2

Greenhouse effect, radiation budget, and clouds

CATHERINE GAUTIER AND HERVÉ LE TREUT

Behind every cloud is another cloud.
— Judy Garland

À partir de ce jour j' n'ai plus baissé les yeux,
J'ai consacré mon temps à contempler les cieux,
À regarder passer les nues,
À guetter les stratus, à lorgner les nimbus,
À faire les yeux doux aux moindres cumulus....
—Georges Brassens, *L'Orage*, 1960

Introduction to climate system energetics

The climate system and its main components (oceans, atmosphere, land, and biosphere) are responding to two main forcings: the heating resulting from the absorption of solar radiation and the dynamic forcing associated with the rotation of the Earth. Other energy sources, like the radioactive energy from the interior of the planet (geothermal energy), have almost no influence on the climate evolution at the time scales discussed in this book. Measuring, understanding, and monitoring how the solar energy is being used within our environment are therefore essential to determine how the climate may fluctuate in response to both natural and anthropogenic causes.

> **Hervé Le Treut** is a former student of the École Normale Supérieure in Paris, and graduated from the Université Pierre et Marie Curie in 1985. He is a senior scientist at the Centre National de la Recherche Scientifique (CNRS) and has been working on climate modeling within the Laboratoire de Météorologie Dynamique, of which he is currently Director. He has been involved in the IPCC as lead author (2001) and convening lead author (2006). He is also a professor at École Polytechnique and a member of the French Academy of Sciences.

The Earth receives its energy from the Sun, but it is not passive. It radiates energy back toward space; every natural body in the universe with a temperature above absolute 0 °K radiates energy. Climate results from an energy balance between absorbed and re-radiated energy. This balance is expressed as the radiative budget – a key measure of the Earth's energetics. Human activities have the potential to modify the components of this budget, directly or indirectly, and the maintenance of and changes in the Earth radiation budget are the foci of the present chapter.

Basic mechanisms

Although it is about 150 million km from the Earth, the Sun provides our planet with a very large amount of energy. Averaged over a year and over the entire Earth, about 341 Watts of energy for every square m (per m²) reaches the top of the Earth's atmosphere. About 70% of this solar energy (or about 240 Wm^{-2}) remains in the Earth's system, while the rest (about 30%) is reflected back to space. The energy remaining is absorbed by the Earth's main constituents (water, air, soil, vegetation), and thus heats them. Most of the 30% of solar energy that is reflected is due to clouds and small particles in the atmosphere called aerosols. The balance is reflected off bright surfaces like snow and ice or bare soils (deserts) and to a lesser extent, vegetation or water (see Figure 1.1). Without a cooling mechanism, the Earth would be continuously heated, leading to a higher global average temperature; this cooling mechanism is the terrestrial emission of radiation (also called longwave or infrared radiation). Everything on Earth (water, land, atmospheric gases, clouds, and aerosol) emits longwave (infrared) radiation with an intensity that is a function of its temperature; the higher the temperature, the more intense the longwave radiation emitted. The intensity of the radiation emitted depends on the emitting object's temperature raised to the fourth power. The Earth's temperature adjusts itself to maintain a balance between the heating resulting from the absorption of solar radiation and the cooling resulting from the terrestrial emission of longwave radiation. If changes occur in either one of these radiation components as a result of solar input variations, an increase in atmospheric (greenhouse) gases or an increase in aerosols due to a volcanic eruption, the Earth system will warm up or cool down until it reaches a new equilibrium in which the heating and cooling are again in balance. This equilibrium may be achieved by means of a change in the Earth's temperature, a change in the distribution of the global cloudiness, or some other natural response of the system. In fact, any perturbation of the energy input and output to the Earth system triggers a complex series of responses and feedbacks, which combine to achieve a new equilibrium. This chapter is intended to shed some light on these processes.

How do we know the Earth's radiation budget?

Radiation observations have a long history, going back to the scientists who worked out some of the key concepts described later in this chapter: the greenhouse effect (the French Joseph Fourier in 1824 [7]) and the role of anthropogenic CO_2 (the Swedish Svante Arrhenius in 1896 [6]). However, systematic observational studies are more recent. The first extensive cloud climatology was compiled by London in 1957 [8]. He assembled a large number of surface cloud observations from the Northern Hemisphere recorded in the 1930s and the 1940s and separated them according to cloud types. This cloud climatology allowed scientists to compute the global radiative budget using a model that computes the amount of radiation reflected, emitted, and absorbed by the different components of the Earth (called a radiative transfer model). By computing the Earth's radiative budget, London found that the Earth's albedo was about 30%, very close to the value of 29% later obtained with more accurate methods (e.g. satellite observations). Since then, many other surface-based climatologies have been compiled.

However, with the advent of space observations, a more global way to observe the radiation budget and its components is satellite observations, which offer a quasi-instantaneous view of the Earth. Such satellite observations started in the early 1970s [9] and are being used relatively routinely, with sensors becoming more sophisticated and data reduction and processing more elaborate, to produce a highly accurate net radiation budget on a global scale. These satellite sensors measure the radiation from the Sun (also called broadband shortwave radiation) reflected and longwave radiation emitted by the Earth, in addition to the total energy emitted by the Sun. From shortwave radiation reflected, the albedo of Earth, which is the fraction of the global incident solar radiation that is reflected back to space, can be determined. For instance, the Earth Radiation Budget Experiment (ERBE) satellite, launched in 1983 [10], provided the most accurate estimation of the Earth radiation budget. ERBE scanned the Earth with longwave, shortwave and total energy detectors. It took 10 years for the final Earth radiation data products from ERBE to be released, but they were the first accurate set of data available to answer scientific questions such as, What is the impact of clouds on climate? These data also helped to tune climate models by comparing their top-of-atmosphere (TOA) radiation budget to ERBE observations (see Figure 2.1).

In Europe, a similar system, the ScaRaB radiometer, was launched in 1994. It consisted of parallel telescopes with different filters for visible, solar, and total radiation, as well as one for the atmospheric window that is the spectral region transparent to the infrared radiation [11]. Later, similar instruments (e.g. GERB) have been launched on geostationary satellites to provide high temporal

Greenhouse effect, radiation budget, and clouds

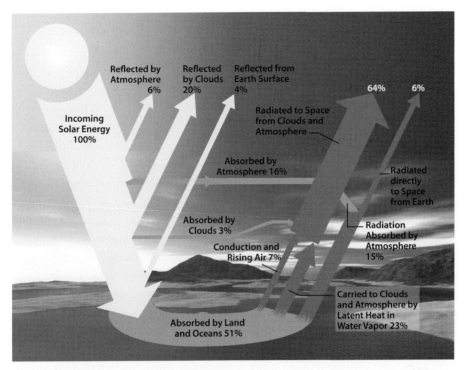

Figure 2.1 Earth radiation budget.

resolution regional observations. More recently, the Cloud and Earth's Radiant Energy System (CERES) has been launched on several satellites and is now providing continuous monitoring of the Earth's radiation budget.

The Earth radiative forcing, and its impact on atmospheric and oceanic motions

All processes that alter the radiative budget of the Earth have an impact on the climate system. The notion of 'radiative forcing' has been introduced to distinguish the role of different components that are likely to be modified by natural processes (clouds) or result from human activities (greenhouse gases, aerosols). An inherent difficulty with this diagnostic is that it depends on the state of the atmosphere; all contributions become mixed up as the climate system begins to evolve as a result of a particular forcing. To solve this problem, and following a definition that has been agreed on by the scientific community, radiative forcing is defined as the modification of the Earth's radiative budget caused by a particular factor, before the Earth's surface and the overlying troposphere are able to evolve, while assuming that higher up in the atmosphere, the stratosphere remains in radiative equilibrium. In practice, the radiative forcing is a metric: it cannot be observed and needs to be computed.

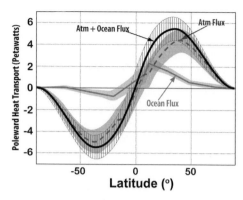

Figure 2.2 Poleward heat transport by the ocean and atmosphere and the combined value as diagnosed by the Pilot Ocean Model Intercomparison Project. The two fluids together carry heat from its region of input in the tropics to high latitudes where the atmosphere radiates it back to space. The combined system is the true 'conveyor' where the ocean dominates in low northern latitudes but is generally a weaker transporter of heat than is the atmosphere.

A positive radiative forcing tends, on average, to warm the Earth, while a negative forcing tends, on average, to cool it. The notion of radiative forcing is also very important because it offers a way to determine the individual contribution of various components (e.g. water vapor, greenhouse gases, aerosols, clouds, snow) to the Earth's radiation budget.

The solar radiative forcing is positive (more heating by shortwave solar radiation absorption than cooling by terrestrial emission of longwave radiation) in the tropical regions and negative (more cooling than heating) in the polar regions. The contrast between the two is the primary mechanism that drives the atmospheric and oceanic circulations, because this radiative imbalance must be balanced by meridional heat transport by the atmosphere and oceans (at least on a long-term basis in which no energy storage and no change in the global mean temperature occur). This means that heat must be exported from the equatorial regions toward the poles. Using observations of the meridional distribution of radiation, the oceans have been found to transport about one-third of the heat from the tropics toward the poles, with a global maximum transport occurring around a latitude of about 20° (see Figure 2.2). The atmosphere also transports heat poleward through extratropical weather systems, but its maximum transport is displaced toward the poles and located around a latitude of about 35°.

This radiative impact on ocean and atmosphere dynamics is the origin of a variety of other processes. As the upper ocean is heated by the incoming solar radiation, heat and water are transferred from the ocean to the atmosphere through turbulent motions. The water evaporated from the ocean is lifted by its own buoyancy, by boundary layer turbulence, or by convective clouds, with motions organized on a hierarchy of scales ranging from km to planetary scale oscillations. Convective clouds are responsible for transporting heat and momentum from the Earth's surface to the upper troposphere. In the tropics, this overall upward vertical motion represents the upward branch of a meridionally oriented cell-like circulation called the 'Hadley Cell' that has its descending branch around latitudes of about 30°, over the desert regions of the world.

These desert areas tend to stabilize the climate of the Earth because they are characterized by a negative radiative forcing, due to a high surface albedo, warm surface temperatures, and a dry and cloud-free atmosphere. This overall radiative cooling is maintained by subsidence (descending motion) warming, which also has a drying effect and therefore helps to maintain the desert characteristics. Another cell-like circulation called the 'Walker cell' exists in the zonal (east-west) direction and is associated with the zonal radiation distribution. Its ascending branches are located in convective regions over warm surfaces (e.g. Western Pacific, Amazon) and descending branches over drier regions.

In higher latitudes, the impact of solar radiative forcing is quite different, as the surfaces are cold and often covered by snow and ice. Most of the radiation in these regions is reflected by the high albedo surfaces. The heat that is absorbed serves to melt snow and ice, and only a small amount remains to heat the surface. As a result, the atmosphere is stratified (with relatively cold, dense air near the surface), and little vertical motion occurs. This means that the effect of any climate change at these latitudes will be felt most strongly near the surface.

The contrast between land and oceans also has a major influence on how radiation affects atmospheric circulation. In the Southern Hemisphere, this contrast is mainly meridional; the radiation distribution is such that the lines of constant net radiation budget are parallel to the lines of constant latitude. The middle and high latitudes of the Northern Hemisphere display an obvious land-ocean contrast at any given latitude, as well as a strong seasonal modulation.

Greenhouse gases and their effects on climate

Whereas the 'greenhouse effect' concept is often associated with the idea of anthropogenic impacts on climate, the greenhouse effect is primarily a natural process, present on other planets in the solar system as well as on Earth. Here, it has a significant and mostly positive impact on the living conditions at the surface. It is also one of the subtlest and most sensitive processes involved in the climatic equilibrium of the Earth. This sensitivity explains why the greenhouse effect can be strongly affected by human activities.

The atmosphere is largely composed of oxygen (21% of its mass) and nitrogen (78%). The remaining part, less than 1%, consists of other gases and condensed particles. Some of these other gases, despite their very small concentrations in the atmosphere, exert a very strong influence on climate. In particular, they can absorb part of the infrared radiation emitted by the Earth's surface. The absorbed energy is reradiated back toward the surface (increasing ground temperature) and partly toward space. This process is called 'greenhouse effect' because it is similar to the effect caused by a thin layer of glass in a greenhouse; solar radiation penetrates through the glass, but the infrared radiation emitted back is not allowed to escape freely, causing warming. A similar process occurs within a car left in the sun: the sunlight heats the interior of the car (e.g. the seats absorb the solar radiation), which then radiates more energy as its temperature increases. The infrared radiation is trapped inside the car, because it cannot escape through the windows. One can easily understand that more opaque windows would result in an even higher air temperature inside the car. Of course, this is only an analogy, as the processes occurring within the atmosphere have many different and unique characteristics (the heating that takes place in an actual greenhouse or in a car is mostly due to the fact that the air is contained in a small space and cannot mix by convection). However, overall, the greenhouse effect acts by reducing the efficiency of the Earth's cooling mechanism through longwave emission. A surface temperature of only $-18\,°C$ would emit the 240 Wm^{-2} of solar energy absorbed by the system and thus would be the equilibrium temperature without greenhouse gases. The atmospheric absorption by the greenhouse gases present in the atmosphere forces the Earth's surface to warm beyond that value in order to produce the same cooling to space. The observed average surface temperature is then about $15\,°C$ ($59\,°F$) and would even be higher without the existence of convective adjustments of the temperature profile (i.e. vertical convective mixing that brings cooler air down and warmer air up). In fact, the greenhouse effect essentially acts as though the $-18\,°C$ ($0\,°F$) surface emitting to space were displaced higher up in the atmosphere, at the so-called 'equivalent emission level,' which is generally a few kilometers high. The relative contribution of atmospheric gases to this greenhouse effect is greatly disproportionate to their concentrations. The oxygen and nitrogen have no greenhouse effect (because the laws of physics require that molecules have at least three atoms for a gas to absorb infrared radiation and thus be a greenhouse gas). The greenhouse gas with the largest greenhouse effect is water vapor (H_2O), with 60% of the total effect, whereas it constitutes only 0.2% of the atmospheric mass. It is followed by carbon dioxide (CO_2), with 26% of the effects for about 0.05% of the atmospheric mass, and ozone (O_3), with 8% of the effects, for about 0.02% of the atmospheric mass). Methane (CH_4) and nitrous oxide (N_2O) represent most of the remaining contribution (6%). Carbon dioxide dominates the effect of all other gases after water vapor. This is surprising, because its radiative forcing per molecule is relatively small, varying with the logarithm of the concentration, because the effect of CO_2 saturates at concentrations less than 100 ppm, and any additional CO_2 will only mildly impact the radiative forcing. Its total concentration, however, is almost 100 times larger than that of any other greenhouse gas and therefore contributes to its large impact.

The effects of these greenhouse gases also differ from each other because of the large disparity in their 'residence times' (the length of time a gas remains in the atmosphere before it is either absorbed by vegetation or the ocean or destroyed through some chemical reaction). The residence time has a direct consequence on how human activities may affect greenhouse gases' atmospheric concentrations. For instance, the recycling of water vapor occurs in a few weeks, and therefore, human activities have essentially no direct impact on the hydrological cycle. On the other hand, the situation for CO_2 is quite different. When it is added to the atmosphere, about three quarters of the extra CO_2 is eventually removed by dissolution in the ocean over a period of about 200 years, while the remainder is removed by chemical reactions with $CaCO_3$ or igneous rocks on land or in the ocean, this last stage taking up to tens of thousands of years. Thus, for the purpose of climate predictions, it is usually assumed that CO_2 lifetime is on the order of 100–200 years. The CO_2 released by the burning of oil, natural gas, or coal can therefore accumulate within the atmosphere over decades or centuries (or at least partially, since half of the emitted CO_2 is captured very quickly by the oceans or the continental vegetation), strongly modifying the atmospheric CO_2 concentration. The growth of other greenhouse gases' concentrations associated with human activities is detailed below.

Anthropogenic increases in greenhouse gases lead to perturbations of the Earth's radiation budget, which may, at first glance, seem very modest; the current impact of the anthropogenic climate effect on the Earth's radiation budget is estimated to be about 2.9 Wm^{-2}, hardly more than 1% of the absorbed solar energy. Regardless, the temperature of the Earth is 300 °K, and a sustained perturbation of the energy cycle at the percent level can produce a global warming of a few degrees, a situation that has not occurred in more than millions of years. The anthropogenic greenhouse effect is therefore potentially damaging, because it disrupts a sensitively adjusted balance between radiative processes, a balance that critically depends on very small amounts of greenhouse gases.

The role of clouds

One of the main unknowns regarding the effects of radiation on climate relates to the role of clouds in heating or cooling the Earth. Clouds form and dissipate through a wide range of processes, from the microphysical processes responsible for droplet formation to the air dynamics at different scales. Clouds themselves have been a source of mysteries for many centuries: why does water not fall out of the sky? Descartes imagined that clouds were constituted of bubbles. We know now that they are made of droplets small enough (less than 15 μm) to be subjected to a strong effect of air viscosity, which ties their motions to that of the atmosphere. When droplets grow beyond the 15-μm limit, they fall as rainwater. Cloud droplets generally do not form spontaneously. When the atmospheric water vapor concentration in the atmosphere reaches near-saturation, the formation of cloud droplets generally requires the help of particles or gases that serve as cloud condensation nuclei (CCN). Whereas different types of aerosols (natural like sea salt or anthropogenic like sulfates) can serve as cloud condensation nuclei, all aerosols are not CCN. The subsequent life cycle of the droplet is also complex; as many more vapor molecules condense on the initially small cloud droplet, it grows. It also grows through interaction with other cloud droplets through a process called accretion.

The net radiative forcing of clouds is the balance between their effect on the shortwave solar radiation within the climate system and on the longwave terrestrial radiation emitted to space. Both effects are important. Clouds reflect solar radiation, thus reducing the amount of solar radiation available for heating the Earth system; this represents a cooling effect (negative cloud radiative forcing). Clouds also act as greenhouse gases do, absorbing the longwave radiation emitted by the Earth surface and radiating energy out to space and downward to the surface; this represents a warming effect (positive cloud radiation forcing). The sum of these two effects depends on the physical characteristics of the clouds.

While cloud observations have been made from the ground for decades and compiled to estimate the global radiative budget for over a century, it is only with the advent of satellite observations that we have had sufficient confidence in the data to convincingly evaluate the role of today's clouds on the climate system. After many controversies in the 1980s over whether clouds warmed or cooled the climate, observations from ERBE settled the issue by accurately determining that, overall, clouds produce a negative radiative forcing of about -13 to -21 Wm^{-2}: they cool the climate by that amount [12]. Many more satellite observations have confirmed this important result in the last decade and have provided for continuous monitoring of the global radiation budget.

For the most part, the effects of clouds on the climate are complex; they are felt at the top of the atmosphere, at the surface of the Earth, and within the atmosphere. They depend on cloud horizontal extent ('cloud cover') as well as on a great variety of cloud properties. The 'cloud optical thickness' – a combination of the physical cloud thickness and the density and nature of cloud droplets or cloud crystals – quantifies the interactions within the shortwave (solar) range. The larger the cloud thickness, the more reflection and the less transmission occur. Clouds also reflect radiation back to the surface, creating multiple reflections between them and the surface. Another cloud parameter that affects the radiation budget is the cloud droplet size distribution, usually summarized by a term called the 'effective radius.' Smaller droplets reflect slightly more than larger ones, because a greater number of small droplets has more reflective surface area than the same water volume composed of fewer but larger drops. This cloud droplet size effect is smaller than that of the optical depth; cloud effects on solar radiation are more influenced by the overall number of droplets than by their size. The 'cloud droplet effective size' varies with the surface over which the droplets are formed because they depend on the presence of condensation nuclei (CCN) in the air from which the cloud is formed. When more aerosols are present, more CCN are available upon which water can condense and more droplets can form. With higher concentrations of aerosols, more of the water vapor molecules available for condensation are used to create new droplets rather than to increase the size of existing droplets. The end result is a smaller effective droplet size (which means that a preponderance of smaller droplets in size distribution of the droplets) than for a cloud embedded in a 'cleaner' atmosphere. Therefore, clouds within which the same amount of water is distributed over a larger number of smaller droplets cover a larger surface, reflect more solar radiation, and cool the climate system. This is known as the 'first indirect aerosol effect' or the 'Twomey effect,' named for the scientist who first suggested this mechanism (see Chapter 3 on aerosols). Another consequence of this process that leads to a preponderance of smaller droplets is that with an equivalent amount of liquid water available, less precipitation will form; droplets have more difficulty reaching the critical size of about 15 μm, beyond which they can fall almost freely in the form of rain.

The phase of water within clouds also plays a role in their effect on radiation. Water droplets reflect more radiation than do ice particles. Ice clouds, like cirrus, which are located at high altitudes, have a small shortwave radiation effect. However, they have a large effect on longwave radiation because the main parameter governing cloud impact on longwave emission radiation is temperature (and therefore cloud altitude, due to the high correlation between altitude and temperature in the atmosphere). In the lower part of the atmosphere (the troposphere), where clouds mostly are located, the temperature decreases with altitude. Therefore, as the altitude of a cloud increases, its longwave emission decreases and its radiative forcing increases. Thus, high clouds like cirrus generally have a strong greenhouse effect and warm the atmosphere below. This greenhouse effect of clouds is commonly experienced at night when cloudy winter nights are much warmer than very clear ones. Even in the daytime, as cirrus clouds are usually thin and made of ice, they have little impact on the shortwave radiation budget and therefore have a minimal cooling effect. Overall, then, the radiative forcing of cirrus clouds is positive; they warm the atmosphere. Middle clouds, on the other hand, are thermally neutral, inducing neither warming nor cooling. As a consequence, changes in cloud altitude distribution will constitute a major factor affecting changing climates.

A changing world: the increased forcing of greenhouse gases

The radiatively important greenhouse gases, introduced in the previous section – carbon dioxide (CO_2), methane (CH_4), chlorofluorocarbons (CFCs), N_2O, HFCs, PFCs and SF_6, ozone (O_3) and anthropogenic water vapor (H_2O) – exert a radiative forcing on climate that is currently increasing, the magnitude of which has been carefully estimated by the Intergovernmental Panel on Climate Change (IPCC) Fourth Assessment Report [5].

The present concentration of CO_2, of about 380 ppm is associated with an estimated radiative forcing of about 1.63 W m^{-2} compared with preindustrial levels. The increased heating is felt at the surface and within the troposphere, whereas the stratosphere cools because the added CO_2 increases the equivalent emission level. The atmospheric concentration of CO_2 has increased globally by about 100 ppm over the last 200 years from its preindustrial value of about 280 ppm to almost 380 ppm in 2004. The CO_2 concentration growth has substantially accelerated over the last 50 years, with a sustained growth rate of nearly 2 ppm yr^{-1} over the period 2001–03. Emissions from fossil fuel burning, cement production, and gas flaring have increased rapidly from 6.1 to 6.5 Gt C yr^{-1}, or 0.7% per year. This rate exceeds the rate that was predicted for the business-as-usual (BAU) scenario in the IPCC Third Assessment Report [3]. Because we know that all the CO_2 emitted by human activities is not found in the atmosphere, a crucial question to address in order to better predict what might happen in the future is how much CO_2 each component of the Earth system takes up and by which mechanisms. This issue is discussed in Chapter 8 on the carbon cycle.

The second-largest anthropogenic radiative forcing after carbon dioxide is due to methane (CH_4). Prior to the industrial age, methane concentration varied from 400–700 ppb but has since increased to a present concentration of about 1,778 ppb. Present atmospheric levels of methane are unprecedented in the last 500 000 years for which we have records. It is estimated that the increase in radiative forcing due to methane since the preindustrial concentration of 715 ppb is about 0.48 ± 0.05 Wm^{-2}. This represents a forcing about four times smaller than that of carbon dioxide despite the fact that, on a per-molecule basis, methane is about

21 times more radiatively active than carbon dioxide. Over the last 25 years, the abundance of methane has increased by about 40%, but its growth rate has decreased substantially over that period. The reason for that decrease is unknown, as are the implications of that decrease for the future. Both are the subject of ongoing research.

The next gas in order of radiative forcing impact is CFC-12, one of the gases regulated by the Montreal protocol. Other gases also regulated and continuously monitored are CFCs, HFCs, chlorocarbons, bromocarbons, and halons, which are all radiatively active gases but with smaller radiative forcings at present.

Nitrous oxide (N_2O) comes in fourth, with a concentration of 314 ppb, an increase of 44 ppb over preindustrial levels of 270 ppb. It has a radiative forcing of 0.15 Wm^{-2}. The primary driver for the N_2O increase is thought to be the enhanced microbial production due to expanding fertilized lands.

Human-synthesized PFCs, HFCs, and SF_6 can have a very significant radiative forcing, even at small concentrations. Their concentrations are increasing rapidly and their extremely long lifetimes (3200 years in the case of SF_6) are a significant concern.

The radiative forcing of ozone (O_3) has two components. The first one, a consequence of ozone depletion in the stratosphere, results in a negative radiative forcing. This forcing varies seasonally, as discussed in Chapter 11. The other, resulting from ozone increase in the troposphere, is mostly due to vehicle emissions and causes a positive forcing.

Finally, it is difficult to differentiate anthropogenic from naturally produced water vapor in the troposphere, but it is clear that most of the stratospheric water vapor is due to human activities and linked to methane increase. Stratospheric water vapor is expected to have a radiative forcing of about 0.02 Wm^{-2} and will increase. Further details on changes in atmospheric gases are provided in Chapter 11 on atmospheric chemistry.

All these forcings vary over different time scales. For instance, over interannual time scales, studies suggest a connection between tropical greenhouse trapping and the state of the tropical atmosphere characterized by the interannual El Niño Southern Oscillation (ENSO). A distinct increase in tropical-mean greenhouse trapping of $\sim$2 Wm^{-2} was observed in the 1986–87 period in conjunction with a $\sim$ 0.4 °K increase in tropical mean sea surface temperature [13]. Also, while it long has been assumed that the effect of long-term variations in greenhouse gas concentrations may have a negligible effect on short time scale variability of the climate system, some studies now suggest that seasonal variability could also be affected by slowly varying CO_2 concentration trends [14].

A summary of all the radiative forcings discussed above is provided in Figure 2.3.

42 Catherine Gautier and Hervé Le Treut

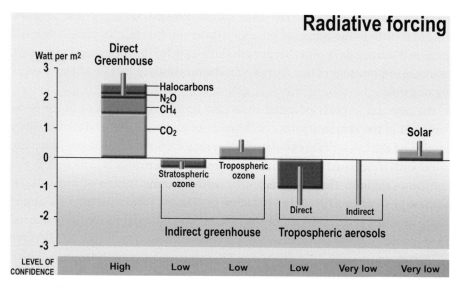

Figure 2.3 Global, annual mean radiative forcings (Watt/m^2) due to various agents for the period from preindustrial (1750) to late 1990s. The vertical bar indicates an estimate of the uncertainty range (after IPCC SAR [2]).

Feedbacks resulting from greenhouse warming

Whereas the increase of greenhouse gases within the atmosphere constitutes a direct impact of human activities on the Earth's energy balance, their role can be affected by important 'feedback effects.' Without such feedback effects, the global surface temperature increase associated with a doubling of the atmospheric CO_2 is about 1.0 °C (1.8 °F) at equilibrium based on theoretical computations. Yet the different numerical climate models presently available provide an estimate from slightly less than 2.0 °C (3.6 °F) to slightly more than 5.0 °C (8.9 °F) for a doubling of atmospheric CO_2. The difference from the 1 °C reference mentioned above is a measure of the feedback effects. The range of estimations indicates that there are remaining model uncertainties associated with those feedbacks.

Some of the feedbacks are directly linked to the greenhouse mechanism itself. Indeed, as discussed before, the natural greenhouse effect manifests itself by an elevation of the effective radiating level (the approximate level where temperature is about −18 °C) by up to a few km. The anthropogenic greenhouse effect amplifies this upward displacement. How such a modification affects the surface temperature at the ground level strongly depends on how the vertical temperature lapse rate (change in temperature with height) is modified. The temperature lapse rate is partly controlled by the occurrence of convection, which mixes the air column vertically when its stratification is unstable; i.e. when the vertical decrease of temperature with height is larger than a given threshold. The convective mixing provides a strong coupling between atmospheric temperature changes at different levels throughout the troposphere – and this is correctly reproduced in climate models. However, other processes may be at play and modify the lapse rate. Whether the lapse rates simulated by climate models are fully representative of what occurs in the real atmosphere is still a debate. This debate reached a very acute intensity when tropospheric temperature records seemed to demonstrate that mid-tropospheric temperatures over the last decades were decreasing rather than increasing as the surface temperatures increased [15]. These temperature records were based on measurements from the Microwave Sounding Unit on the NOAA satellites. The measurements raised concerns about models' abilities to properly represent the correlation between temperatures at different atmospheric levels. The problem turned out to be largely due to an incorrect data interpretation, mostly an inadequate treatment of the signal contamination by stratospheric cooling (an effect which is associated with tropospheric warming, when greenhouse gases increase, as seen earlier), and other technical issues. This finding put an end to the debate concerning observations of temperature changes at different levels in the atmosphere and reinforced our confidence in climate models' abilities to correctly handle the atmospheric temperature lapse rate. Of course, this does not provide an absolute validation of the models; uncertainties concerning the evolution of the temperature lapse rate in a warmer world still affect the predicted surface temperature at different latitudes.

Perhaps the most important atmospheric feedback is due to water vapor and is linked directly to the greenhouse effect itself. Indeed, water vapor is the main natural greenhouse gas, accounting by itself for between 36% and 66% of the greenhouse effect and in the presence of clouds between 66% and 85% of it. As mentioned above, water vapor content cannot be affected directly by human activities; however, the Earth's warming will modify water vapor distribution through two key processes. As the atmosphere gets warmer, the saturation threshold for water vapor concentration is increased and the atmosphere can hold more water vapor. Also, the water vapor concentration is generally increased through increased evaporation, further amplifying the warming. In most climate models, this water vapor feedback doubles the climate response to any perturbation and therefore is crucial for future climate predictions. In fact, the capacity of existing models to adequately represent it is a continuous issue of concern, and the successive IPCC reports have given increasing attention to uncertainties related to the amplitude of water vapor feedback. Recently, however, modeling results for the special atmospheric temperature conditions following the Mount Pinatubo volcanic eruptions have permitted an assessment of the modeled water vapor feedback. The strong agreement obtained between observed and modeled temperature reduction and decreased water vapor (integrated over the atmosphere and in the upper troposphere) about 3 years after the eruption (i.e., with plenty of time for the water vapor to equilibrate with the cooler sea surface temperature) suggests that climate models can simulate quite adequately the water vapor feedback, although some uncertainties linger. For instance, recent observations in the tropics suggest that tropospheric water vapor does not fully adjust to constant relative humidity, as usually assumed in models, and a definite confirmation will be impossible to obtain. Nevertheless, many other indices suggest that models are, overall, broadly correct, because water vapor feedback is essential to reproduce adequately the temperature and climate fluctuations at a variety of time scales (seasonal, interannual, paleoclimatic, etc.).

Climate-induced changes in the surface albedo constitute another major feedback. Surface albedo changes modify the warming by solar radiation at the surface, with the most important example associated with the melting of ice and snow in a warmer world. In most model studies in response to a CO_2 doubling, this feedback increases the warming by about 0.5°. It is also known that this so-called 'snow/ice albedo' feedback played an important role at paleoclimatic time scales; for example, during transitions from glacial to interglacial epochs. However, as with the water /vapor feedback, its accurate estimation with models is difficult. Correctly evaluating surface albedo changes when snow melts under a heavy forest, for instance, or taking into account the effect of snow aging on its reflective properties, are still major challenges for climate models.

This list of feedbacks is not comprehensive. For example, vegetation changes linked to soil desertification or even land-use changes also modify the Earth's surface albedo and give rise to a 'vegetation albedo' feedback. Other feedback effects may be operating through a modification of the surface roughness and the consequent changes in turbulent energy exchanges between surface and atmosphere. All of these feedback processes induce large uncertainties in climate model performances. Still, the largest source of complexity and uncertainty is certainly the effect of clouds, the extreme complexity of which has been stressed earlier. As mentioned above, human beings also have the potential to influence cloud formation through the production of aerosols by industrial activities or land-use change and to affect cloud lifetime, since the presence of aerosols reduces the probability of rain formation. Instead of raining out, clouds containing high concentrations of aerosols remain in the atmosphere longer, creating an additional overall cooling effect. More details on this effect will be provided in Chapter 3 on aerosols.

Finally, it is important to stress that the above description does not include all aspects of the climate machinery that interact with the greenhouse forcing. Moist and dry areas over the globe are determined by the location of the Hadley and Walker cells, which themselves are the result of forcings other than radiative processes, such as the rotation of the Earth, the pole-to-equator temperature gradient, or the land-ocean distribution. Climate warming may, in turn, affect these large-scale atmospheric circulations, thereby causing changes in the location of deserts, humid areas, and their associated radiation balances.

Priority issues for research: models

Fully assessing the role of the greenhouse effect on climate in the coming decades will require a drastic improvement in climate models and climate model validation.

The models currently used were developed in the late 1970s and have continuously improved since then. For example, in the early 1980s, most models did not predict cloudiness but used observed climatologies based on London's data compilation to compute radiation fluxes. Over the years, models have incorporated a much more comprehensive representation of clouds. This representation was first based on the use of simple predictors such as relative humidity and vertical stability, but later made use of a complete equation for the prediction of cloud water, including its generation (through condensation) and dissipation (through precipitation and evaporation) processes.

Despite this continuing progress, climate models' uncertainty in terms of estimating climate sensitivity has changed little over the last few years. For example, the computed temperature response to a CO_2 doubling still exhibits a range of about 2–5° between different models, largely the result of cloud effects. Some models exhibit a strong positive feedback associated, for example, with the tendency to replace low clouds with higher clouds in the CO_2-rich future (with, therefore, a stronger greenhouse effect). The introduction of cloud microphysics in the model has suggested a possible influence of negative feedbacks associated with the increase of cloud water content (especially when ice clouds are replaced by water clouds). The continuous improvement of models and consequent increased complexity has therefore not helped to reduce the range of warming predicted by the different models. This is not surprising in view of the complex parameters and processes that characterize cloud interaction with radiation. Whether some of these processes are strong enough to be dominant and dictate clouds' response or whether model-limited predictability reflects a true unpredictability of the real world is still an open question. Furthermore, models remain considerably simpler than the real world, and cloud representations are still heavily parameterized – that is, not explicitly computed but represented by parameters based on observations – because most of the dynamical processes at the origin of cloud formation have a much smaller scale than the 100 km of a model grid box. The limited size of clouds is made obvious by simply looking at them through the window.

However, the situation is rapidly changing. Some attempts to insert simplified (two-dimensional) cloud-resolving models within each global model grid-box [16] and the development of global models with resolution of a few km are both indications that we are on the verge of major transformations in climate modeling. These latest models have yet to be run for long simulations of climate change or used to explicitly address the feedback problems mentioned in this chapter. When they are, not all problems will be eliminated. Recent studies suggest that part of the reason for the discrepancies between current models may be due to the differing representations of low stratiform clouds, which still require a parameterization approach. The advances discussed above will nevertheless undoubtedly affect our perspective concerning water vapor and cloud feedbacks.

Monitoring Earth's radiative budget during the current climate change

The advent of new satellite instruments also offers an important opportunity to advance our understanding of the role of radiative processes in the context of climate change.

Whereas only a few of the cloud parameters affecting radiation could be unambiguously derived from satellite measurements, new instruments being put into orbit will modify our perspective. Space-born lidars (CALIPSO), radars (CLOUDSAT), instruments to measure directional reflected and polarized solar radiation (PARASOL), and very high resolution spectrometers (AIRS, IASI) associated with more conventional instruments are already beginning to offer a wealth of information concerning cloud top height, cloud phase (water or ice), cloud microphysics (droplet or crystal size), and cloud water content (see Figure 2.4). Efficiently using this considerable amount of information will require the development of new and adapted analysis methods. Some studies have shown the potential of using in-depth analysis to carry out meaningful comparisons with model simulations. Another potential approach is to directly estimate the increasing global warming signal while comprehensively monitoring other climatic properties.

Observations of the radiation budget components are ongoing from both surface and space. Despite the short length of time over which satellite observations have been made and the difficulty of obtaining a global picture from the longer land-based time series, some preliminary trends are emerging, the interpretation of which are still challenging. In particular, a controversy has emerged regarding the observed variations of solar radiation at the surface and the Earth albedo over the last two decades. A decline in solar radiation over land surfaces has become apparent in many observational records up to 1990, a phenomenon called 'global dimming' (see, for example [17]). However, newly available surface observations

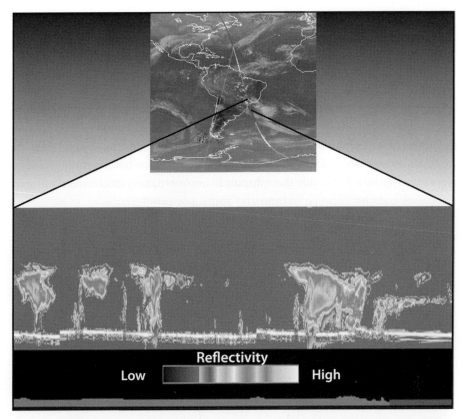

Figure 2.4 Clouds at CLOUDSAT satellite trajectory overlaid on geographic map (top); cross-section of radar data along CLOUDSAT satellite trajectory over the Andrea tropical storm on May 9, 2007, indicating cloud characteristics expressed in reflectivity (bottom). High values of reflectivity correspond to high water content. (For image in color, please see Plate 2.)

from 1990 to the present, primarily from the Northern Hemisphere, show that the dimming did not persist into the 1990s; instead a widespread brightening has been observed since the late 1980s. Measurements of Earth's albedo from space made with CERES since the 1970s indicate that the global annual Earth albedo (whose measured average value is about 0.29) has slightly decreased (by about 0.006) over this period [18]. This corresponds to a decrease of about 1 Wm^{-2} in shortwave reflected flux, and therefore a radiative warming of the Earth-atmosphere system by that amount. A similar decrease in Earth's albedo has been reported from lunar observations [19]. Did the decrease in Earth's albedo and increase in global surface radiation result from decreased global cloudiness? Satellite longwave radiation observations do not support this hypothesis [18]. A clue about the origin of the brightening comes from surface observations that show the brightening to be occurring under both all-cloud and clear-sky conditions, indicating that processes in both cloud-free and cloudy atmospheres contribute to this brightening. This suggests a possible interplay between direct and indirect aerosol effects.

These preliminary results and the associated controversy point to the need for uninterrupted and highly accurate global radiation measurements. Such measurements, combined with concomitant continuous observations of complementary parameters (e.g. cloud and aerosol parameters), will allow us to fully capture changes as they occur and unravel the processes responsible for these changes. This example also illustrates the complex interplay that exists between different processes and also between natural and anthropogenic forcings.

Conclusions

The increasing atmospheric concentration of greenhouse gases is probably the most important and largely undisputable evidence with which the scientific community has begun to alert citizens and policy makers about the risk of a climate change. The mechanisms through which this forcing may disrupt the current climatic equilibrium are well understood through models and verified through observations. Unfortunately, our capacity to diagnose the existence of these climatic risks is not yet met by a capacity to carry out fully quantitative predictions of the future climate. This is because of the many complex processes and feedbacks that determine, in particular, the modifications of clouds or their interactions with aerosols.

Whether the uncertainty related to these feedbacks, and particularly to the cloud feedback, will be reduced during the coming decades, when a new generation of higher resolution models becomes available, a wider spectrum of climate parameters are continuously monitored, and climate change signal becomes easier to diagnose and understand, is a key element in the relations between scientists and society.

3

Atmospheric aerosols and climate change

YORAM J. KAUFMAN AND OLIVIER BOUCHER

We cannot solve our problems with the same thinking we used when we created them.

– Albert Einstein

N'oublie pas que chaque nuage, si noir soit-il, a toujours une face ensoleillée, tournée vers le ciel.

– Friedrich Wilhelm Weber

A variety of aerosol types

Atmospheric aerosols are liquid and solid particles suspended in the atmosphere. This definition excludes cloud droplets, ice crystals, and other hydrometeors, which are discussed in Chapter 4. Atmospheric aerosols encompass a large range of sizes, from a few nanometers to a few tens of nanomers, and exhibit a large variety of chemical composition. Individual aerosol particles cannot be seen

Yoram J. Kaufman (1948–2006) received his PhD from Tel-Aviv University, Israel, in 1979. He was a senior fellow and atmospheric scientist at NASA's Goddard Space Flight Center, Greenbelt, Maryland. He served as project scientist of the Earth Observing System Terra satellite. He was a fellow of the American Meteorological Society and the 2006 recipient of the Vernon E. Suomi Award in recognition of his highly significant technical achievements in the atmospheric sciences. Kaufman is listed among the Institute of Scientific Information (ISI) highly cited scientists in geophysics.

Olivier Boucher is a former student of the École Normale Supérieure in Paris and graduated from the Université Pierre et Marie Curie in 1995. He is a senior scientist at the Centre National de la Recherche Scientifique (CNRS), currently on secondment at the Met Office Hadley Centre in the United Kingdom, where he is head of the Climate, Chemistry and Ecosystems team.

with the eyes due to their small size, but their collective effect can be seen on solar radiation when their concentration is elevated. Indeed, we have all experienced a hazy day because of a dust storm, sea spray, or smoke from vegetation fires or industrial activity. Aerosols are emitted in the atmosphere from a number of sources, and as a consequence, different types of atmospheric aerosols exist. Some aerosols are natural, such as sea spray and volcanic ash, while others are due to human activities, such as smoke from deforestation and industrial haze. Moreover, some natural sources are magnified by human activities, such as dust emissions from arable land and road traffic.

We usually distinguish primary from secondary aerosols. Primary aerosols are produced either from combustion sources or mechanically, from the friction of wind on the ocean or land surface. Examples of primary aerosols include desert dust, industrial dust, sea spray, smoke from forest fires or biomass burning associated with deforestation, and some agricultural practices. Domestic use of biofuels such as wood or dung is also a significant source of primary aerosols in many developing countries. By contrast, secondary aerosols originate from gaseous species, which then condense to form particles in the atmosphere. Sulfate aerosols, which have both natural and anthropogenic sources, form the most important type of secondary aerosols. Most aerosols in the lower part of the atmosphere have distributions characterized by high regional maxima due to their short lifetime of about a week and the regional nature of the sources. Meteorological conditions determine how far aerosols are transported from their source regions as well as how they are vertically distributed through the atmosphere. Aerosol concentrations and properties are modified during transport by dry or wet deposition, in-cloud processes, and atmospheric chemical reactions. Figure 3.1 illustrates the large variability in aerosol types and properties.

Aerosols from regional pollution are mainly fine submicron hygroscopic particles found downwind of populated regions in air polluted by car engines, industry, cooking, fireplaces, etc. The very same aerosols are also responsible for acid rain and health hazards such as air pollutants. Environmental regulations have contributed to decrease the aerosol concentrations in the last decades by up to a factor of two in Europe and North America. However, aerosol concentrations increased in the developing world in tandem with population and economic growth. The main ingredients of the industrial aerosol are sulfate, which stems from oxidation of sulfur dioxide, and black carbon, which stems from incomplete combustion.

Smoke from vegetation fires is dominated by submicron organic particles with varying amounts of black carbon emitted in the hot, flaming stage of the fire. In forest fires, the flaming stage is followed by a long, cooler, smoldering stage during which wood emits large amounts of smoke composed mainly of organics with little black carbon. Conversely, African grasses burn quickly in strong flaming

Atmospheric aerosols and climate change 51

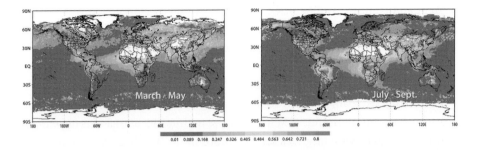

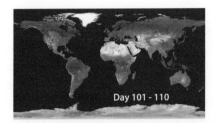 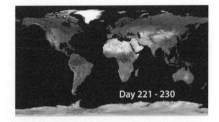

Figure 3.1 Global distribution of aerosols and vegetation fires for March–May 2005 (left) and for July–September 2005 (right). Regions with open biomass burning appear in red on the lower panels and are characterized by elevated aerosol optical thickness. Other aerosol types, such as dust blowing out of the Sahara, sea-salt aerosols over the oceans and pollution aerosols over industrialized regions, are also visible on the upper panels. All the data are from the MODIS instrument on the NASA Terra satellite. The fire information is from http://rapidfire.sci.gsfc.nasa.gov/firemaps, and the aerosol information from http://lake.nascom.nasa.gov/movas/. (For image in color, please see Plate 3.)

fires, emitting large quantities of black carbon, without a smoldering stage. Dense smoke plumes are found every year from fires over extensive areas in South America (August–October), Central America (April–May) Southern Africa (July–September), and Central Africa (January–March).

Desert dust, mainly large mineral particles of up to several micrometers in size, is emitted from dry lake beds in the Sahara, East Asia, and the Saudi Arabian deserts. Very little dust is observed over Australia, where the topography is mostly flat, because the arid regions are old and highly weathered. An unknown amount of dust is emitted from disturbed soils in Africa and East Asia. One estimate is that 10% of the global dust originates from human activity. Not surprisingly, dust production is largest in drought conditions, such as during the strongest El Niño year of 1983. African dust is transported across the Mediterranean Sea to Europe and across the Atlantic Ocean to Florida, where during the summer months it occasionally causes local pollution to exceed the US Environmental Protection Agency standards on particulate matter (PM). Dust from East Asia, from natural sources and land use, can be lifted up to an altitude of 3–5 km and transported during April and May to North America. Intercontinental transport is associated with dust deposition to the Atlantic and Pacific Oceans, providing a key nutrient – iron – to oceanic phytoplankton.

Oceanic aerosols are composed mostly of micron-sized salt particles emitted from bursting sea foam in windy conditions and submicron sulfate particles from emissions of dimethylsulfide (DMS) by phytoplankton. The emission is dependent on the wind speed with large sea salt emissions in the 'roaring 40s' – the latitude belt around 40 °S.

Like greenhouse gases, aerosols have a profound effect on our environment and climate. However, there is a fundamental difference in how greenhouse gases and aerosols operate because of their very different atmospheric lifetimes. Long-lived greenhouse gases with lifetimes of 10–100 years and homogeneous spatial distributions dominate the global and long-term effects of human activity on climate, whereas tropospheric aerosols with lifetimes of a week have played a significant role in the climate of the industrial period and generally exhibit more regional features. It is crucial to quantify the interplay between the radiative effects of greenhouse gases and aerosols to understand both regional and global climate change. This requires continuous aerosol observations from satellites, networks of ground-based instruments, and dedicated field experiments. These observations can, in turn, constrain and feed global climate models.

The known, the less known and the unknown

Climate varies naturally but can also be forced to respond to systematic changes in atmospheric composition. The radiative effects resulting from an increase in the concentration of anthropogenic aerosol or greenhouse gases, radiative forcings discussed in Chapter 2, cause a change in the Earth's radiative balance, which is the basic ingredient of climate change. A positive radiative forcing indicates that the Earth-atmosphere system gains radiative energy, resulting in warming. Models of the climate system show that the global surface temperature rises by 0.4–1.2 °C (0.7–2.1 °F) for every 1 Wm^{-2} of additional radiative forcing. There are several pathways in which aerosols exert radiative forcing, which we now discuss in association with Figure 3.2.

Aerosols scatter and absorb solar radiation and, to a lesser degree, thermal radiation emitted by the Earth's surface and the atmosphere. By scattering and reflecting sunlight back to space, aerosols reduce the amount of sunlight reaching the surface and therefore cool the Earth system. This direct effect of aerosols on climate is now well documented and understood. Anthropogenic aerosols have contributed to limit global warming over the industrial period, but with a

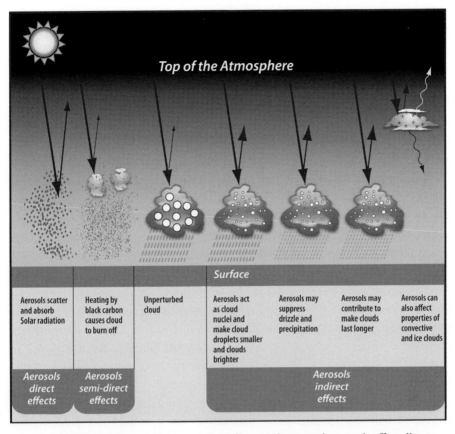

Figure 3.2 Schematic representation of how anthropogenic aerosols affect climate.

present-day radiative forcing estimated to be in the range of −0.2 to −0.8 Wm^{-2}, their importance remains uncertain. Absorbing aerosols, such as soot, are thought to play a particularly important role. These aerosols redistribute solar energy by heating the atmosphere and cooling the surface below. Although their net effect on the overall Earth's radiative budget may be small, absorbing aerosols have a much larger effect on the surface energy budget. Decreasing the amount of solar radiation reaching the surface can suppress the evaporation flux, thus slowing down the water cycle. Absorbing aerosols may also affect cloud formation and precipitation. For example, measurements over the Amazon forest show that clouds essentially disappear in the dry season when smoke is present. However, under some conditions, aerosol absorption can also strengthen low-level clouds. Moreover, the deposition of soot on snow and ice darkens the surface in polar regions, which can exacerbate the snow albedo feedback and further warm these regions. All of these aerosol effects nevertheless remain very uncertain.

Each cloud drop requires an aerosol particle on which to condense. Without aerosols, clouds would not form, or they would do so under much higher atmospheric moisture content. Thus, the concentration, size, and composition of the aerosols, which can act as cloud condensation nuclei (CCN), play an important role in controlling the cloud properties and the development of precipitation. Cloud formation, however, is mainly driven by the availability of moisture and other dynamic processes. Although natural aerosols are needed to form clouds, it seems that industrial haze and smoke aerosols take every opportunity to reduce the formation of precipitation (as though intentionally trying to reduce their washout from the atmosphere!). Aircraft measurements show that a six-fold increase in the concentration of fine aerosols produces a three- to five-fold increase in the cloud droplet concentration. Analysis of global satellite data shows that such a change in aerosol concentration corresponds with 10%–25 % smaller cloud droplets, because the condensed water is split into more numerous droplets. Clouds with smaller, more numerous droplets have a larger surface area and, therefore, a larger reflectivity by up to 30%. This increase in the reflection of sunlight to space was proposed three decades ago to possibly rival the greenhouse gas forcing. If the cloud modifications observed globally were attributed to anthropogenic effects solely, they would translate into a present-day additional radiative forcing of −0.5 to −1.5 Wm^{-2}. This aerosol indirect effect is still very uncertain and subject to continuous research.

Not only does the aerosol-induced increase in cloud droplet concentration contribute to radiative forcing, but it also affects the dynamics and precipitation processes in clouds. In clean conditions, the cloud droplet size increases as the cloud develops and extends in the vertical direction until the droplet reaches a critical radius of about 10–15 micrometers for the onset of liquid precipitation. Alternatively, the droplet may freeze if the cloud top temperature reaches −10 °C (14 °F). In polluted air masses, satellite data show not only smaller droplets at the cloud base, but also a failure of cloud droplet size to increase as the cloud develops. Consequently, precipitation does not occur or is delayed in polluted water clouds. This suppression of precipitation was also observed in stratocumulus clouds polluted by emissions from ship stacks and in polluted cumulus clouds over the Indian Ocean. The high aerosol concentration supplies new condensation nuclei on which to condense the excess water vapor as the cloud cools down. This results in an increase in the cloud liquid water content, lifetime, and extent. Liquid water clouds that cannot precipitate due to the high concentration of aerosols could still precipitate once the droplets freeze. However, measurements in deep convective clouds show that the freezing process is postponed until the cloud reaches cold temperatures of −37 °C (−35 °F). Strong updrafts and condensation together lead to a high concentration of small cloud droplets. These droplets do not collide efficiently to form raindrops, nor do they freeze at −10 °C (14 °F). This effect results in supercooled liquid water, thus eliminating an alternative way for clouds to precipitate. The global importance of this effect is still not clear.

The aerosol-induced reduction in the initial droplet size at the cloud base, reduction of growth of the droplets as the cloud develops, and postponement of freezing at lower temperatures are processes that work toward making precipitation more difficult to form until new (cleaner) conditions are met downwind and dynamic processes overcome the aerosol effect, with precipitation washing out the aerosols in the process. Analysis of a 50-year record of precipitation in California and Israel shows precipitation losses over topographical barriers downwind of major coastal urban areas that amount to 15%–25% of the annual precipitation. The precipitation suppression occurs mainly in relatively shallow orographic clouds. The suppression that occurs over the upstream slope is compensated by a similar percentage enhancement on the drier downstream slope.

One can therefore wonder whether aerosols can change the regional amount of precipitation, as redistribution does not necessarily change the total amount of precipitation. There are actually two mechanisms by which aerosols can reduce total precipitation. The first is by reducing surface solar radiation, thereby reducing evaporation from the oceans and land, which will directly affect the availability of water vapor for precipitation. The second mechanism is through changes in the atmospheric circulation. Two modeling studies showed that aerosol absorption of sunlight over land in East Asia and over the Indian subcontinent can enhance the circulation from the sea to the land, increasing water vapor availability and precipitation.

In summary, man-made aerosols supply more numerous cloud condensation nuclei and reduce cloud droplet size. These effects trigger an array of side effects and feedbacks that – depending on the cloud type, topography, geographic location, and local air circulations – can increase or decrease cloud cover and therefore the amount and distribution of precipitation. Man-made aerosols are like a dangerous genie let out of the bottle.

Measuring aerosols, an international collaborative effort

Measuring atmospheric aerosols is a major challenge. Unlike gases, atmospheric aerosols cannot be solely characterized by their concentration. The full characterization of atmospheric aerosols requires knowledge about their size distribution (how many particles in each size range), chemical composition, shape, state of chemical mixture, and optical properties. Additionally, we need to understand how aerosol properties vary with humidity.

Obviously, we are still far from a comprehensive characterization of aerosols, although decades of research have seen enormous progress in measurement techniques. One can broadly classify measurement techniques in two categories, *in situ* and remote-sensing measurements. *In situ* measurements consist of aerosols sampled in the atmosphere and then taken to the laboratory for analysis, whereas remote sensing measurements operate remotely and exploit the interaction of aerosols with electromagnetic radiation to infer aerosol properties. Most of the *in situ* measurements are performed from the Earth's surface, but aircraft or balloon platforms offer useful alternatives. Remote sensing measurements are usually performed from the Earth's surface and from space using satellites. They can occasionally be made from aircraft and balloon platforms. Both types of measurements, *in situ* and remote sensing, have their advantages and disadvantages. *In situ* measurements can measure a large range of aerosol physical, chemical, and optical properties. However, most *in situ* measurements require manipulating the aerosols in one way or another. This can potentially alter the physical and chemical properties of the aerosols, especially their water content. Remote sensing measurements are appealing because they measure the ambient, undisturbed aerosol and can focus on aerosol optical properties, which are important for the climate system. Moreover, a global coverage of the Earth is only achievable through satellite remote sensing measurements. Satellite measurements have drawbacks, too, as they are currently limited to a few of the many parameters that would be needed to understand the interaction between aerosols and climate. Yet capabilities of satellite missions are not exhausted, and future missions will extend our current capability.

Ground-based aerosol measurements

By measuring transmitted and scattered solar radiation from the ground, it is possible to retrieve accurate aerosol information, such as the column aerosol scattering, absorption, and size distribution. The AERONET (AErosol RObotic NETwork[1]) program is a federation of ground-based aerosol remote sensing networks, originally established by collaboration between NASA and the French space agency CNES (through the French PHOTONS network) and further extended through the Canadian AEROCAN network. AERONET's goal is to monitor aerosol optical properties and validate satellite aerosol retrievals. With over 200 instruments, the network successfully imposed standardization of instruments, calibration, and processing with identical instruments owned by national agencies, research

---

[1] See http://aeronet.gsfc.nasa.gov/.

institutes, and universities. Two basic radiometric measurements (of attenuated sunlight and sky brightness) made in eight spectral bands are transmitted from instruments all over the world to three geosynchronous satellites and then retransmitted to the appropriate ground receiving station. The network provides globally distributed near–real-time observations of aerosol spectral optical depths, size distributions, and column water vapor in diverse aerosol regimes. This international collaboration with strong United States and French components has improved tremendously our understanding of atmospheric aerosols.

Space-based aerosol measurements

However, even this impressive AERONET network is not enough to provide the data coverage scientists need to understand the large heterogeneity of aerosol properties worldwide. Thus, space-based Earth observing satellites are necessary. With their variety of Earth orbits (sun-synchronous or geostationary), satellite sensors are excellent tools for studying and monitoring atmospheric aerosols. Radiant energy reflected and emitted by the Earth shows signature of the atmospheric and surface properties. By measuring the spectral, angular, and polarization properties of this radiation, satellite sensors can quantify several atmospheric and surface properties, including aerosols.

Remote sensors can be understood by analogy to human eyes, which are sensitive to a narrow range of radiation in the visible spectrum from violet to red light. Humans have depth perception because our eyes have slightly different angles of observation. Aerosol remote sensing was first developed using a single spectral channel and a single angle of observations (corresponding to a color-blind person with one eye). With the advent of the TOMS instruments, with two spectral channels sensitive to ultraviolet (UV) light, observations of elevated smoke or dust layers above a scattering atmosphere were possible. However, the first instrument designed specifically for aerosol measurements was POLDER, introduced in 1996. It used a combination of spectral channels covering a range wider than human vision and a wide-angle camera that observes the same target on Earth at several different angles. Additionally, POLDER measured light polarization to detect fine aerosols over the land, taking advantage of the difference between the spectrally neutral polarized light reflected from the Earth's surface and the spectrally decreasing polarized light reflected by fine aerosols. In addition to POLDER, two other instruments (MODIS and MISR) on the Terra satellite have been measuring global aerosol concentrations and properties since 2000. Over the ocean, MODIS uses the aerosol spectral signature in a wide range of the visible and near-infrared wavelengths to distinguish small pollution particles from coarse sea salt and dust. Over land, aerosol properties are more difficult to retrieve, and MODIS

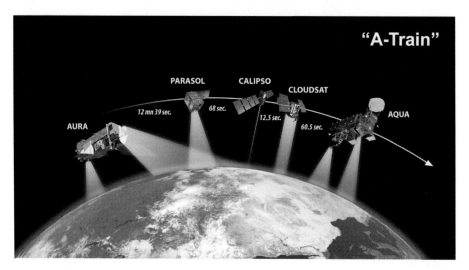

Figure 3.3 'A-Train': A constellation of satellites flying in close formation in order to sample quasi-simultaneously the state of the atmosphere. The various instruments measure aerosols, clouds, some gas-phase chemical species, and relevant atmospheric parameters, such as the vertical profile of moisture and temperature. PARASOL is a CNES mission, Calipso is a CNES–NASA mission, and other satellites are NASA missions.

uses a combination of channels to differentiate surface from aerosol properties. MISR detects the reflected light at different viewing angles along the satellite track, which provides information to separate the aerosol signal from that of the surface.

The recent realization that aerosols affect surface temperature, cloudiness, and precipitation patterns created a demand for more informative space-borne observations. A major international collaboration between NASA, CNES, and the European and other space agencies led to the deployment of the 'A-train,' a series of satellites flying in formation with several satellite instruments designed to measure aerosols, clouds, and other relevant atmospheric properties (Figure 3.3). In addition to MODIS on AQUA and POLDER on PARASOL, the OMI instrument on AURA measures the aerosol effect on UV radiation, which is sensitive to the aerosol absorption in the UV and dust outbreaks over the desert surfaces. Complementing these observations, CERES on AQUA measures the reflected solar radiation and the emission of thermal radiation to space; CALIPSO and CLOUDSAT, launched on April 28, 2006, measure the vertical distribution of aerosol and cloud liquid water; and AIRS and AMSU measure profiles of atmospheric temperature and water vapor. It is anticipated that synergistic use of these instruments will help to resolve some of the current unknowns about aerosols and their interactions with clouds and climate.

Aerosols and societal issues

Despite their important role in climate and climate change, atmospheric aerosols have not received as much attention from the media and the public as have long-lived greenhouse gases. The situation is changing, however, and there is now an increasing awareness of the roles that aerosols play in the Earth's climate system. At the same time, the contribution of aerosols to a range of environmental problems has been largely documented for several decades now. Sulfate aerosols are responsible for acid rain, with negative effects on sensitive ecosystems, such as boreal forests in North America and Europe. The role of aerosols – known to air quality managers as PM – in air pollution is being increasingly recognized at local and regional scales. Because of their short residence time in the atmosphere, aerosols are found in higher concentrations closer to their source regions. Consequently, industrial emissions of aerosols are highly relevant to all three issues of acid rain, air quality, and climate change.

The impact of industrial sulfur dioxide emissions on acidification became clear in the early 1980s, and policy decisions to cut sulfur dioxide emissions were implemented in North America and Europe during the 1980s and 1990s. This has resulted in significant reductions in emissions that translated into a major decrease in the concentration of sulfate aerosols, as evident from observations. As epidemiological studies demonstrate with increasing clarity that a link exists between atmospheric concentrations of PM and a number of illnesses such as asthma or cardiovascular diseases, legislations on PM atmospheric levels and emissions in developed countries are becoming more stringent. We are therefore now in a situation in which aerosol emissions and concentrations recently have been reduced in North America and Europe, at least for some of the chemical species composing the aerosols. Because of their different economic trajectories, emissions of aerosols and aerosol precursors in developing countries are still increasing. Projection of aerosol emissions in the future is a difficult exercise, but it is anticipated that industrial emissions will continue to decrease in developed countries. In developing countries, it is likely that emissions of aerosols will also decrease in the future but perhaps not until the standard of living and attention to health issues increase in these countries. Ironically, such a global reduction in aerosol emissions, which is essential for the good of human health, will contribute to further warming of the regions where aerosol emissions will be decreased. This is nevertheless difficult to quantify precisely, because the climate response to a radiative forcing can be both regional and global. Numerical simulations show that a complete shutdown of current industrial aerosol emissions would result in an extra warming of about 1 °C (1.8 °F), which would appear gradually within a few decades. This remains an uncertain estimate; because the current cooling by aerosols is uncertain, the potential warming associated with the suppression of this cooling effect is equally uncertain.

It is unfortunate that aerosols have such detrimental effects on human health and ecosystems and interfere with precipitation processes, because they have the potential to offset some of the warming induced by anthropogenic greenhouse gases. Some experts are promoting the use of aerosols as part of a climate change engineering strategy. If aerosols are harmful in populated regions of the Earth's surface, why not introduce them over the ocean or high up in the atmosphere? One such engineering solution consists of injecting bright aerosols into the stratosphere (an altitude of 10 km or higher). Because aerosol removal processes are very slow in the stratosphere, the residence time of aerosols there is roughly several years. By reflecting solar radiation back to space, these 'artificial' stratospheric aerosols would decrease the amount of solar energy absorbed by the planet and could, as a consequence, slow or even counteract global warming. Others have proposed an artificial mechanical production of sea salt aerosols at the ocean surface, with a similar effect on the radiation balance of the Earth.

As this chapter and this book illustrate, many uncertainties remain in our knowledge of the functioning of Earth's climate system. It is therefore difficult to predict the climate's response to such voluntary radiative forcings. For instance, it is not clear how high-level cirrus clouds would respond to increased stratospheric aerosol loadings as aerosols progressively fall through the atmosphere. Also, low-level clouds would respond to increased production of sea salt aerosol in a very uncertain way, with possible feedbacks on the atmospheric general circulation. Along with uncertain and possibly unknown climate responses to a voluntary human aerosol forcing, there could also be undesirable side effects or impacts that would need to be balanced against the potential benefits on surface temperature. Typically, increased stratospheric aerosol loadings would affect the incoming solar radiation at the surface, with further impacts on plant productivity. Moreover, the surface radiation budget also governs the evaporation rate, with impacts on the water cycle and precipitation. In a complex and interdependent system such as the climate system, interfering with one element may feed back on many other elements. In other words, there is great danger in further interfering with a climate system that we do not understand fully. Additionally, because of the short lifetime of aerosols compared with long-lived greenhouse gases, aerosol geoengineering would have to be sustained over many human generations to offset the greenhouse gas emissions of one human generation. Within our current scientific and socioeconomic knowledge, aerosol geoengineering does not appear to be a sensible answer to the climate change issue.

Atmospheric aerosols have multiple effects on the Earth's climate. Anthropogenic emissions of aerosols from biomass and fossil fuel burning play a key role in explaining the evolution of the Earth's temperature since the beginning of the industrial period. However, many uncertainties remain, in particular with respect to the role of natural and anthropogenic aerosols on clouds and precipitation. The growing recognition of the importance of aerosols effects on climate, the water cycle, and human health will continue to foster research in this area. Our understanding of atmospheric aerosols will improve as new observational and modeling tools become available. If we do not understand aerosols sufficiently, we will never properly deal with short-term climate change.

Recommended books and articles

Intergovernmental Panel on Climate Change, *Climate Change 2001, The Scientific Basis. Contribution of Working Group I to the Third Assessment Report of the Intergovernmental Panel on Climate Change.* (Cambridge: Cambridge University Press, 2001), chap. 5, 6.

Intergovernmental Panel on Climate Change, Climate Change 2007: *The Physical Science Basis: Contributions of the Working Group I to the Fourth Assessment Report of the International Panel on Climate Change.* (Cambridge UK/NY: Cambridge University Press, 2007).

Y. J. Kaufman, D. Tanré and O. Boucher, A satellite view of aerosols in the climate system. *Nature*, **419** (2002), 215–23.

V. Ramanathan, P. J. Crutzen, J. T. Kiehl and D. Rosenfeld, Aerosols, climate, and the hydrological cycle. *Science*, **294** (2001), 2119–24.

O. Boucher, L'influence climatique des aérosols. *La Météorologie, 8ème série,* **17** (1997), 11–22.

4

The global water cycle and climate

MOUSTAFA T. CHAHINE AND PIERRE MOREL

> I've lived in good climate, and it bores the hell out of me. I like weather rather than climate.
>
> – John Steinbeck

> Un modèle sans observation n'est qu'un exercice mathématique gratuit. Des observations sans modèle n'apportent que confusion
>
> – Jacques-Louis Lions

Introduction

We call our planet Earth, clearly because we live on land and see ourselves surrounded by solid ground. Yet by right, it should be called 'planet Ocean' because the ubiquitous presence of water is the feature that uniquely

Moustafa T. Chahine's primary interests are remote sensing of planetary atmospheres and surfaces and studies of climate change processes. He joined the Jet Propulsion Laboratory (JPL), California Institute of Technology, in 1960, where he served as the founding head of the Division of Earth and Space Sciences and as the JPL chief scientist until 2001. Currently, he is Science Team Leader for NASA's Atmospheric Infrared Sounder (AIRS), which was launched on board the Aqua spacecraft in 2002. Aqua is part of NASA's Earth Observing System and is designed to study Earth's water cycle and energy fluxes.

Pierre Morel's fields of interest have included atmospheric circulation dynamics, climate science, and Earth observation from space. In his capacity as Professor at the University of Paris, he created the Laboratory for Dynamic Meteorology (LMD) in 1968. He was principal investigator for the French–American EOLE satellite project and the initiator of the European meteorological satellite program METEOSAT. He also served as Deputy Director-General of the French Space Agency (CNES) (1975–1982), Director of the International World Climate Research Program (1982–1994), and Scientific Advisor for the NASA Earth Science Enterprise (1995–2001).

characterizes our planet among all celestial bodies in the solar system and makes it shine like a lone blue jewel in the immensity of space.[1] Indeed, the ocean was the cradle of life on the planet and remains a haven for innumerable living species, while the availability of freshwater in rivers and in the ground is essential for sustaining life over land. Water in the atmosphere plays an even more basic role in determining the planetary environment and making it suitable for the existence of substantial biological activity. As a matter of fact, one may foresee that this felicitous state of affairs will not last forever. The planet is steadily losing water – actually hydrogen atoms – to space and getting insufficient replacement from icy meteorites and other extraterrestrial sources. This inexorable process will eventually (after about a billion years) make the Earth, like Venus, a parched desert roasted by the searing greenhouse effect of accumulating carbon dioxide. Until this extreme fate befalls the planet, water vapor, condensed water and ice clouds, and rainfall will be major components of the climate system and will affect the planetary environment in many crucial ways, as discussed here and in other chapters of the book.

Water in the atmosphere fixes the current equilibrium point of the Earth climate and, to a large extent, controls the climate response to changes in external boundary conditions or factors. Conversely, the atmospheric circulation and weather drive the cycling of water among the atmosphere, oceans, and land. For the purpose of this book, it is convenient to consider the atmospheric water cycle in relation to climate and climate change as a specific issue. This leaves the impact on the ocean circulation – changes in ocean salinity and density due to local imbalances between precipitation and evaporation – to be discussed in Chapter 6. Likewise, the partitioning of rainfall between evaporation back to atmosphere, infiltration into ground reservoirs, and run-off to streams and freshwater bodies, is covered in Chapter 5, along with the crucial issue of future water resource availability.

The condensation of atmospheric water vapor and subsequent rainfall release vast amounts of energy that feed the 'atmospheric engine' and weather. Active weather systems, such as mid-latitude traveling disturbances or tropical hurricanes and typhoons, are just huge heat engines fueled by the latent heat of condensation of water vapor they draw from air. Conversely, weather disturbances, from vast cyclonic systems to medium-scale storms or even single-cell convective clouds, inject water from the surface into the atmosphere and induce water vapor removal through precipitation.

---

[1] After the British novelist Arthur C. Clarke (born 1917): 'How inappropriate to call this planet Earth, when clearly it is Ocean.'

Atmospheric moisture and climate

One may begin by recognizing that water vapor is the atmosphere's main greenhouse gas. On global average, atmospheric water vapor cuts down the loss of radiant energy – infrared radiation – from the planet to space by some 90 Wm^{-2}. This value is to be compared with the aggregate effect of the much-advertised permanent greenhouse gases like carbon dioxide, methane, nitric oxide, ozone, etc., altogether 28 Wm^{-2}. It stands to reason that even a small error in the water vapor budget of the planetary atmosphere or, even more cogently, the adjustment of atmospheric water vapor to changing climate conditions has a major impact on the computed radiative balance. The calculated impact of a 5% increment in predicted relative humidity of the free atmosphere (above the well-mixed boundary layer) is typically 2 to 3 Wm^{-2}, depending upon ambient moisture content.

Relative humidity is the ratio of the number density of water molecules actually present in the air to the maximum allowed number for a particular ambient temperature (see Figure 4.1). As the definition implies, the highest allowed relative humidity is 100%, at which point the excess water must condense and fall out. This maximum water molecule density or water-holding capacity of air rises sharply with increasing temperature, according to the well-established Clausius-Clapeyron law. Any cooling machine, such as an air conditioner, causes the condensation of the excess water that cannot remain in gaseous state within the colder air.

By far, the largest fraction of 'precipitable water' present in the atmosphere lies in the warmest layers, up to about 2 km above the surface, primarily because of the rapid decrease in water-holding capacity of air with decreasing temperature. In fact, water vapor in the near-surface 'planetary boundary layer' is close to saturation and narrowly linked to surface temperature (over land, soil moisture is also a significant factor). This state of affairs is convenient for model designers, because knowledge of air temperature is enough to infer the atmospheric moisture content. Unfortunately, this is insufficient to quantify the loss of radiation to space. The blanketing effect of the atmosphere occurs mainly at altitudes where the air is significantly colder than the ground and grows sharply as the temperature difference increases, so that water molecules in the cold 'upper troposphere' are responsible for a disproportionately large fraction of the greenhouse effect. The small fraction – between 5% and 10% – of total atmospheric moisture above 5 km or 3 miles altitude produces half of the total greenhouse effect of atmospheric water.

Water vapor in that region – the upper troposphere – is far from saturation with respect to either liquid water or ice, so that the local water-holding capacity is not the controlling factor: upper tropospheric water vapor cannot be assessed by

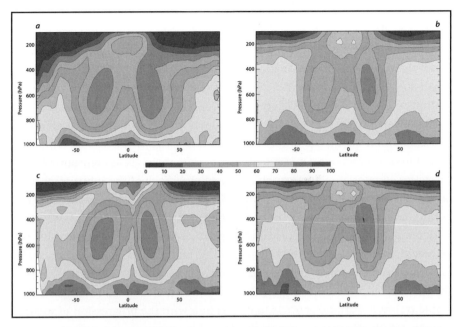

Figure 4.1 Zonal-average relative humidity (%) as observed by the NASA Atmospheric Infrared Sounder (AIRS) and as computed by various climate models for northern winter months: December, January, and February. (a) AIRS observations over the period 2002–06; (b) average of 19 climate model simulations over the period 1980–99; (c) 1980–99 model simulation by Japan's Frontier Research Center for Global Change; (d) 1980–99 model simulation by the United Kingdom Meteorological Office. (Figure is courtesy of NASA and JPL. The panels were produced by Drs. Thomas Hearty and Duane Waliser of JPL. For more information, see http://airs.jpl.nasa.gov.) (For image in color, please see Plate 4.)

ambient temperature only. One must take into account the complex and highly variable processes that govern the water budget of individual air parcels, as evidenced by daily, even hourly variations in atmospheric moisture and net radiation. State-of-the-art climate models are far from being able to adequately represent these phenomena.

Moist processes and clouds in nature

The injection of water vapor from the water-rich boundary layer near the surface into the free atmosphere is accomplished by all sorts of vertical motions, from individual 'convective' cloud cells to continent-scale weather systems. In the tropics, vertical ascent is generally concentrated within narrow 'cumulonimbus' towers that can shoot up to quite high altitudes with considerable velocity. Even modern aircraft are well advised to avoid flying into such clouds, lest they be damaged by the associated wind-shear and aerodynamic turbulence. The descending return flow takes place much more slowly over a relatively large volume surrounding the ascending tower. While such cells serve primarily to dry out the atmosphere through the production of torrential rainfall, they also moisten the upper air by turbulent 'detrainment' from the saturated central tower and 'sublimation' or re-evaporation of the ice cloud particles that stream out from the head of the tower. The ice content of the outflow in turn depends on the efficiency of precipitation within the tower, which itself is the result of various competing microphysical processes. On a much larger scale, moisture is transported from the active equatorial region into the descending branch of the subtropical circulation that corresponds with the planet's desert belt. This feature of the planetary atmosphere is known as the 'Hadley circulation.'

At temperate latitudes, vertical mixing produced by occasional storm cells competes with the broad ascending and descending motions associated with traveling disturbances that account for the succession of fair and foul weather. On average, the outcome is a relatively stable upper-air moisture content, although considerable transient variations occur in connection with the progression of weather systems. Again, the overall process involves phenomena on very different scales, from molecular to microphysical (for example, the role of condensation nuclei on which water molecules can stick and accumulate), to turbulent eddies around and inside clouds, to storm-size 'meso scale' and even traveling planetary waves. Needless to say, the task of quantifying these diverse processes to the accuracy needed for climate projections is daunting. This is especially so because meso-scale storms are largely self-sustaining fluid-dynamic machines, subject to the availability of energy (primarily latent heat of condensation of the water vapor they can draw from the surrounding air) but otherwise free to evolve according to their own dynamic characteristics and life cycles.

A similar analysis applies to the formation of extended layer clouds that significantly affect the Earth's radiation balance. Relatively small amounts of condensed liquid water or ice have a disproportionately large radiative impact on the planet, as clouds reflect a large fraction of incident solar radiation and also contribute substantially to the total greenhouse effect. As seen from space, the almost ubiquitous cloud cover contributes half of the planetary albedo: lacking clouds, earthlight on the moon would be dimmer by a factor of two. The outcome is that clouds cause a global deficit of some 50 W/m^2 in radiant energy absorbed from the Sun. On the other hand, clouds – especially cold cirrus clouds – absorb a notable fraction of the infrared radiation emitted by the Earth surface, thus adding about 30 W/m^2 to the greenhouse effect of absorbing gases.

Even relatively quiescent stratus clouds have their own internal dynamics that govern their life cycles. Air parcels at the base of the cloud layer are heated by upwelling infrared radiation from the surface and tend to rise. Air parcels at the top are cooled by radiation loss to higher levels or to space and tend to sink. Hence, the formation of heat-driven cellular eddies within the cloud layer that are perceived as 'turbulence' by the passengers of commercial airliners that fly through the cloud. As mentioned above, deeper convective cells (cumulonimbus clouds) are themselves self-organizing heat engines embedded in large-scale weather systems. There are no known means to quantity the full range of responses of these complex dynamic phenomena short of explicit simulation of the individual cellular flows and associated microphysical processes. Quantifying radiation transfer among scattered clouds made of a heterogeneous mix of ice particles and water drops is yet another daunting challenge. Any substantial difference in cloud cover, either in the physical nature (liquid water or ice) of cloud particles, or their size spectrum or their vertical distribution, has a significant impact on radiation transfer within the air column and the resulting radiation budget.

Notwithstanding dedicated efforts to model these complex processes and track the amount of condensed ice or liquid water in individual air parcels the size of a model grid element, climate models still fall far short of a realistic representation of actual cloud microphysics and cloud cell dynamics. We may have to wait years to see the fruition of forward-looking approaches now being tested to attack this problem, such as explicit numerical simulation of cloud dynamics and microphysics within selected small samples of the whole atmosphere [16].

Moist processes in climate models

A fundamental obstacle hampering coarse resolution climate models is the nonlinear nature of the relationship between weather and cloud systems on the one hand and the large-scale circulation of the atmosphere on the other. The average response to instantaneous changes in local temperature and moisture is different from the response to changes in the mean temperature and moisture. Climate models are currently unable to resolve individual weather phenomena and cannot explicitly simulate the dynamics of weather systems or accurately predict the true (nonlinear) aggregate impact of short-lived meteorologic disturbances. Even more seriously, climate models operate on average quantities and parameters that have no directly observable counterparts in nature. Under the best of circumstances, model formulations can only be tested by matching global observation statistics with various model 'products' that aggregate the outcome of many interacting elementary processes into a single number.

The modelers themselves pay relatively little attention to bridging the gap between model formulas and the physical reality. However, one should recognize that adequate data in support of the modelers' empirical formulas are seldom available. So-called model studies, based solely on numerical simulation, are commonly used to investigate subtle aspects of climate change, such as projecting changes in soil moisture storage or tagging molecules originating from different sources. Such model studies usually list a collection of 'sub grid scale physics' formulations but rarely discuss how the findings might be affected by the modeling simplifications imposed on the actual elementary phenomena. Only basic physical laws that apply to elementary processes or verifiable scaling laws, when applicable, are guaranteed to remain unaffected by climate change. Independence from climate change cannot be assumed in the case of empirical relationships that have been fitted to match current climatic conditions.

The report of scientific experts consulted by the IPCC recognizes that 'the controlling processes are represented with varying degrees of fidelity in general circulation [climate] models,' but nonetheless, 'the successes of the current models lend some confidence to their results' ([3], p. 427). No definition of 'success' is given for such model simulations that cannot be compared effectively with any past climate event for lack of adequate knowledge of all of the relevant parameters or boundary conditions. Enough uncertainty remains in the whole construct to allow the continued skepticism of some knowledgeable experts (see, for example [20]). Obviously, climate model projections of moist atmospheric processes are not founded exclusively on irrefutable physical properties and laws. They do rely on a number of empirical relationships that may or may not hold up on closer investigation.

Model formulations are, explicitly or implicitly, adjusted or tuned to mimic readily observable aspects of current climate and transient climatic variations. This tuning is a difficult procedure, because all moist processes in the atmosphere are interconnected through their influence on air temperature, moisture, and the large-scale wind field. Any change in the representation of one particular process (for example, introducing a distinction between water and ice-cloud particles) will affect the outcomes of all other operative processes. As a matter of fact, numerical weather forecasters know well that introducing a manifest improvement in the formulation of one particular process will degrade the overall performance of the model unless all the other processes are suitably tuned to the new 'model climate.' Although model simulations cannot be accepted as solidly established scientific evidence, they will nonetheless likely produce reasonable estimates of the quantities for which they have been tuned. Let us investigate this point a little further.

First, we know that changing the composition of the air will change the blanketing effect of the atmosphere – in a way equivalent to adding an extra flux of radiant energy at the top of the troposphere. The standard yardstick in such matters is doubling the concentration of carbon dioxide from the preindustrial value of 280 parts per million. Energy-wise, this is equivalent to adding an extra input of about 4 W/m² globally. The question of climate sensitivity amounts to asking by how much the Earth's surface should warm in order to balance this extra energy input by a corresponding increase ΔF (W/m²) in infrared radiation lost to space. The relation between ΔF and the change ΔT in global-mean surface temperature (°K) is the basic climate sensitivity factor λ resulting from all the adjustments that take place in the atmosphere.

Trends in global-mean climate properties over a number of years are subtle and largely lost in the 'noise' caused by random weather-related fluctuations, to say nothing of measurement uncertainties. However, the same concept can be applied to the large and easily documented seasonal variations over half the Earth, provided one may take each hemisphere as a surrogate for the whole atmosphere when considering the adjustment to a temperature trend.

This idea may sound odd, but it is not unreasonable. Firstly, the Northern and Southern Hemispheres are not so different as regards the large-scale behavior of the atmosphere. Both have similar climate zones from tropical to polar and similar general circulation around the Earth. Second, the moist processes discussed above are associated with relatively small-scale weather systems, so that a representative number of them can be found within each hemisphere at any one time. Finally, the life cycle of these active weather systems is a matter of hours to days, so a period of six months is plenty of time for adjustment to changes in the large-scale environment. Obviously, seasonal cycling in 'hemispheric climate' is not a perfect analog of the Earth's response to long-term climate forcing, but what would a perfect analog be? Certainly half the Earth cannot be taken as a closed system, since there is significant heat loss between the summer and winter hemispheres. Conversely, seasonal ocean warming and cooling reach a relatively shallow depth compared with downward diffusion of a long-term warming trend. However, we are not looking for an exact energy budget, only for fast adjustments in weather, atmospheric moisture, cloudiness, and outgoing longwave radiation that follow changes in surface temperature.

Let us see how the numbers work out. Introducing data obtained from the NASA Langley Research Center and from climatologist Abraham Oort at the Princeton Geophysical Fluid Dynamics Laboratory, the differences between the 'meteorological summer' months (June, July, and August) and winter months (December, January, and February) can be computed readily. The results are shown in the table below.

Does this tell us something about global warming? Yes, because the near equality of the hemispheric sensitivity factors (despite a factor 2 difference in seasonal temperature ranges) is indicative of a quasilinear dependence between outgoing long-wave radiation and mean surface temperature. The same sensitivity may be expected for the response of the total planetary atmosphere to global warming, as long as such warming remains a modest fraction of seasonal cycling. It is no coincidence that the sensitivity factor of the early single-column climate models used to estimate the global response to changes in atmospheric composition (see, e.g., [21]), was close to $\lambda = 0.5\,°\text{K}/\text{Wm}^{-2}$ ([3], p. 354).

	Northern Hemisphere	Southern Hemisphere
ΔF (seasonal)	21 Wm^{-2}	9.0 Wm^{-2}
ΔT (seasonal)	11.7 °K	5.1 °K
$\lambda = \Delta T \Delta F$ (sensitivity)	0.56 °K/Wm^{-2}	0.57 °K/Wm^{-2}

Modern climate simulations are not so easy to compare with observation, because they incorporate many more components of the climate system, which may respond to environmental changes on much longer timescales than the atmosphere alone (also longer than accurate observational records). These extra components add their own feedbacks to the overall global warming response, hence raising the aggregate sensitivity factor beyond that of fast atmospheric adjustments. For example, coupling the atmosphere to the Arctic ice eventually leads to melting and the disappearance of polar ice in the Northern Hemisphere, with attendant increases in air-sea exchanges and the amount of sunlight absorbed by the Arctic Ocean, thus further enhancing the rise of surface temperature. Indeed, the mean sensitivity factor of 15 climate simulations analyzed in the 2001 IPCC report was just short of $1\,°\text{K}/\text{Wm}^{-2}$, which is not inconsistent with a $0.5\,°\text{K}/\text{Wm}^{-2}$ response of the atmosphere alone.

Thus, we have good reason to believe model projections of global warming and direct consequences – such as glacier melting and sea level rise – not because we trust that models are faithful representations of the actual physical processes but because model formulations are steeped in good empirical knowledge of present climate variability. The reliability of model projections of the less direct consequences of climate change, especially changes in weather and rainfall patterns, is another matter altogether.

The water cycle in a changing climate

Precipitation generally comes in two broad varieties: relatively regular and persistent rainfall from 'stratus' layer clouds or heavy bursts of torrential rain from convective clouds. In stratus precipitation, the ascending motion of the air is forced by developing large-scale weather systems or by the underlying orography: even modest hills can induce precipitation under the right conditions. Stratus precipitation dominates outside the tropics, except for summer rainstorms, and shows spatial coherence over distances of 100 km or more. By contrast, convective precipitation results from relatively small-scale but vigorous cellular circulation sustained by latent heat released by water vapor condensation. Such convective cells are the principal source of rain in the tropics and over continents in summer. In many instances, convective cells are embedded in large-scale precipitating layer clouds and account for the bursts of heavier rain.

Advanced climate and weather prediction models explicitly track cloud water and ice variables in individual (grid-sized) air parcels and are reasonably successful in forecasting rain associated with recognizable weather disturbances, although the challenge of predicting the exact rainfall amounts remains a challenge. Quantitative Precipitation Forecast (QPF) is still a research acronym! On the other hand, convective precipitation can only be estimated with empirical formulas based on the predicted large-scale atmospheric circulation and water vapor distribution. This is especially difficult, because the formation of individual convective cells has a hairline trigger that responds to minor differences in airflow or moisture fields. As a result, convective rain forecasts are still rather poor over continents and of unknown value over the oceans. We do know, however, that model predictions of precipitation statistics are very sensitive to the detailed formulation of the model physics. Even a minor modification in one particular process formulation within a global circulation model can significantly alter the predicted rainfall distribution worldwide [22].

Adding these unresolved modeling difficulties to the uncertainties in evaporation from the oceans or land and in atmospheric water vapor transport and mixing, one must conclude that projections of the global water cycle are quite uncertain. This is to say nothing of bold prognostications one may read about future hydrologic climate in a particular region. Not even the sign of the projected regional changes in hydrologic statistics is confidently known, while the overwhelming impact of the increasing demand on water resources is all too easily foreseeable. One may want to keep in mind, however, that climate change could exacerbate already pressing water shortages in many regions of the world (see Chapter 5).

These unresolved problems constitute a fundamental roadblock in the path toward better climate change predictions, because meaningful climate projections depend on the ability to predict changes in the global water balance. For example, the 'thermohaline' circulation of the deep ocean owes its existence to quite small differences in seawater density, which itself depends on water temperature and salinity, the latter being the most significant factor. Ocean salinity, in turn, directly reflects the dilution of seawater by freshwater exchanges with the atmosphere: precipitation minus evaporation. An example of unresolved problems is the possible weakening of Atlantic transport of warm water from the tropics to high latitudes – the famous Gulf Stream – that could result from a freshening of North Atlantic waters (making it impossible for superficial water to sink into the deep ocean). It could be painful to the locals if such a change in Atlantic Ocean circulation resulted in the extension of the Alaskan climate to similar latitudes in Northern Europe (Oslo and Stockholm are within 1° in latitude from Anchorage). In general, one should keep in mind that freshwater fluxes are on a par with heat and radiant energy fluxes in coupling the atmospheric climate with the other reactive components of the Earth system, the oceans, terrestrial hydrology, Greenland and Antarctic ice caps, and, of course, vegetation. No serious long-term projection of global climate can be conducted without a quantitatively correct prediction of changes in the global water cycle.

This is a daunting task, considering the diversity of moist processes, the scattered and transient character of rain events, and the patchiness of local hydrologic properties. We reviewed briefly the modeling difficulties involved in quantitative precipitation forecasts. Almost as daunting is the challenge of measuring rainfall and establishing a reliable benchmark for early diagnostic of possible trends. Fairly dense and well-tended rain-gauge networks exist in economically affluent regions of the world. With care, such networks can provide adequate rainfall statistics over a period of time, at least on relatively flat terrain like the Great Plains of the United States. Yet even in North America, precise estimations of area-averaged rainfall are not available in mountainous regions. In other parts of the world, rain gauges are just too scarce to be helpful. Over the oceans, obviously, no such observation exists. Some rainfall data are obtained from flat islands but, even on the flattest atoll, the temperature contrast between the island or lagoon and the open ocean is sufficient to create a local 'rain climate' that is not representative of oceanic expanses. Indeed, students of history know that ancient Polynesian sailors could navigate and spot the presence of low-lying islands from miles away, thanks to the tall cloud towers that systematically rise over such islands.

The alternative to rain gauge measurements is indirect estimation based on satellite observations, principally the detection of cold cloud heads (associated with strong convection and a downpour of rain) or measurement of cloud water absorption of the microwave radiation emitted by the ocean. The former method provides useful information about the likelihood and location of heavy rain showers but only a rough approximation of total rainfall, using empirical correlation with rain-gauge statistics that are bound to be climate dependent. The latter method measures the amount of condensed water or ice in the air column and, in principle, is more closely representative of the actual precipitation process. Nonetheless, the 'retrieval' procedure for inferring total rainfall from microwave data does involve various cloud models that introduce a degree of arbitrariness.

Furthermore, the real precipitation phenomena are drastically undersampled by existing satellite observing systems; observations from orbiting satellites are repeated too infrequently to catch the development of individual rain events. In addition, the instantaneous field of view of microwave instruments is much wider than individual clouds, so that crucial small-scale concentrations are missed altogether. Finally, no remote satellite observation is yet capable of measuring snowfall.

For these reasons, one may rely on satellite data to identify regional differences and transient rainfall variations from one month to the next, or from one year to the next, but satellite-based estimates of long-term trends are fraught with all kinds of uncertainties and some serious biases. On the other hand, no other data source comes even near to providing the comprehensive global information that could help refining the current generation of water cycle models.

Conclusion

It is unlikely, then, that progress in global rainfall observation or any other large-scale measurement will provide striking new insight in mean precipitation trends or climate-related changes in moist processes. There is really no alternative to developing much more realistic numerical representations of real-life moist atmospheric processes at their proper scales, and testing these models against detailed observations of the time-dependent three-dimensional structure of precipitating clouds. Such models would be instrumental in establishing the linkages between meso-scale weather and cloud-scale physics, on the one hand, and the large-scale general circulation of the atmosphere and climate on the other.

We can see the limits of the current approach to climate prediction, based on heavily parameterized numerical models that embrace a multiplicity of highly variable physical phenomena in a single formula involving space-averaged quantities. Indeed, it is time to take into account the self-sustaining dynamics of individual weather systems within the overall picture of the atmospheric environment. Finally, it is time for climate science to reconnect with meteorological science (see, for example [23]). But one cannot hide the fact that such a shift in scientific outlook is a tall order. It would be a sea change for all players, especially so in America, where the two communities are traditionally separate.

5

Climate change effects on the water cycle over land

ERIC F. WOOD AND KATIA LAVAL

A river seems a magic thing.
A magic, moving, living part of the very Earth itself.

– Laura Gilpin

Et l'eau ce don du ciel,
L'eau qui, couvrant la plaine ou suintant d'une voûte,
S'épand tantôt par flots et tantôt goutte à goutte,
L'eau qui baigne la fleur,
L'eau qui de ses baisers presse la terre avide,
Seule chose ici-bas qui sans vieillir se ride
Et pleure sans douleur.

– Victor Hugo, *Vers datés*

What is the water cycle?

Water is omnipresent on Earth, where it covers over 70% of Earth's surface and is the fundamental constituent for life as we know it. The presence of water

Eric F. Wood is Professor of Civil and Environmental Engineering at Princeton University, New Jersey, where he has taught since 1976. His research areas include hydroclimatology, with an emphasis on modeling the terrestrial water and energy budgets, and hydrological remote sensing and data assimilation. Dr. Wood is a fellow of the American Geophysical Union (AGU) and of the American Meteorological Society, and he has served on many national and international committees related to World Climate Research Programme activities.

Katia Laval is a professor at the Université Pierre et Marie Curie in Paris, where she is the Director of the School of Doctoral Studies in Environment. Her climate research focuses on the interaction between soil, vegetation, and the atmosphere. She is a national correspondent at the French Académie d'Agriculture.

Climate change effects on the water cycle over land 75

in its three states (or phases) – as a solid (snow and ice), a liquid, and water vapor – is a fundamental characteristic of Earth's climate. As water changes states, energy is either absorbed by water (e.g. for evaporation) or released as in the case of water vapor, condensing and forming rainfall. At global scales, this energy exchange help fuel Earth's climate system and links the water cycle to the energy cycle, which obtains its energy from the Sun's radiation.

The water cycle (or hydrological cycle) describes the movement of water among its various storage components and in its different states. The various storage components consist of the amount of water stored in the ocean, the atmosphere (as water vapor), and the land (as soil water and groundwater – in the soil column, and on the land surface in lakes, wetlands, rivers, or in solid form as snow and ice). Table 5.1a shows estimates in these storages, and Table 5.1b shows the water transfers between these storages from the processes of condensation

Table 5.1a. *Amount of water stored in various components of the hydrosphere (after [25])*

Storage	Volume	Percentage (%) of total water	Percentage (%) of total freshwater
Ocean	$1,338 \times 10^6$ km^3	96.5	
Atmosphere	0.013×10^6 km^3	0.001	0.04
Land (total)	48×10^6 km^3	3.46	
Groundwater: total	23.4×10^6 km^3	1.7	
(Freshwater)	(10.5×10^6 km^3)	(0.76)	30.1
Soil moisture	0.016×10^6 km^3	0.001	0.05
Lakes	0.176×10^6 km^3	0.013	
(Freshwater)	(0.091×10^6 km^3)	(0.007)	0.26
Marsh water	0.011×10^6 km^3	0.0008	0.03
Water in rivers	0.002×10^6 km^3	0.0002	0.006
Ice caps and glaciers	24.1×10^6 km^3	1.74	68.7
Water as permafrost	0.30×10^6 km^3	0.022	0.86

Table 5.1b. *Amount of water transferred between the various components of the hydrosphere*

Transfer process	Land-Atmosphere	Land-Ocean	Ocean-Atmosphere
Precipitation	0.099×10^6 km^3/year		0.324×10^6 km^3/year
Evapotranspiration	0.062×10^6 km^3/year		0.361×10^6 km^3/year
Runoff/groundwater		0.037×10^6 km^3/year	

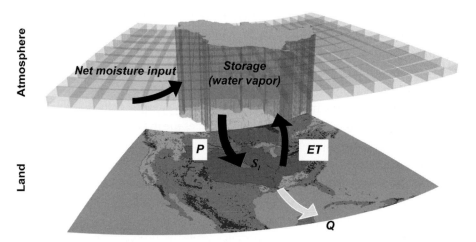

Figure 5.1 Schematic diagram showing water budget coupling between land and atmosphere. Here, the variable ET refers to evapotranspiration, P to precipitation, S1 to land-moisture storage (soils water, lakes, reservoirs, and groundwater), and Q to surface and subsurface runoff. Q links the land with the ocean storage.

(precipitation), evaporation from stored water (lakes, rivers, soil water, snow, and ice), transpiration from vegetation, referred to here as evapotranspiration, runoff, and groundwater discharge into the oceans.

While the term 'water cycle' describes the movement of water within and among these storages, the term 'water budget' (or 'hydrologic budget') is a balance among the transfers over an area (region) and time period, with changes in water storage that include soil water, snow water, glaciers, groundwater, wetlands, lakes, and reservoirs. The net flows into and out of the region include precipitation, both liquid and solid (an inflow); evapotranspiration (an outflow); and discharge out of the region, both as surface runoff (rivers) and as deep groundwater flows (an outflow). This water budget is valid over all spatial and temporal scales, from points to continental areas.

Figure 5.1 shows the water budget of a land area and the overlying atmosphere where, for the land area, the input is precipitation, and the output is the net flow of river and groundwater across the area boundary as well as evapotranspiration. The difference between the inputs and outputs over a time period is the change in water storage (lakes, wetlands, soil water, and groundwater). For the atmosphere overlying the land region, the inputs are the net advection of atmospheric water vapor into the region and evapotranspiration from the land surface (into the atmosphere), while the output is precipitation, and the balance between these is the change in atmospheric water vapor. The major processes of the water cycle are defined more completely in the appendix to this chapter.

Table 5.1a shows that much of the global freshwater is stored in glaciers, ice sheets, and groundwater – storages whose volumes tend to change slowly. The portion of the global freshwater budget that is stored in more dynamic forms (e.g. atmosphere, lakes/reservoirs, and wetlands as soil moisture and in rivers) contributes to substantial variability in the continental water cycle on seasonal, interannual, and decadal timescales. The numbers in Table 5.1b demonstrate that water moves quickly among these dynamically varying storages. For example, the water vapor in the atmosphere is replaced, on average, every 10 days. The intricate connection between the movement and storage of water at the land surface and land-atmosphere processes, and its susceptibility to global climate variability and change, has important implications for society. For example, atmospheric warming increases the water-holding capacity of the atmosphere; this has been supported by recent observations over oceans [24]. If a slowing circulation and convergence do not compensate its effect, this increase leads to enhanced precipitation, either by more intense or more frequent storms – often referred to as 'acceleration of the hydrological cycle' – with increased risk of flooding that seems to be supported by observations.

The remaining sections in this chapter include our assessment of recent change to the terrestrial water budget, based on both observations and modeling studies and a summary of modeling consensus from the Fourth IPCC assessment report [5] carried out for this article, followed by some background material on the challenges in measuring and modeling terrestrial water budget components. These challenges include difficulties in projecting change to variables like evapotranspiration, a process that responds to atmospheric, soil, and plant characteristics and conditions, and in providing attribution to change that could come from anthropogenic climate effects (e.g. CO_2) or anthropogenic land management effects (land use, land cover change).

Observed and modeled changes in the terrestrial water budget components

There is consensus in the hydrology and climate community related to current trends and projection of climate variables relevant to the terrestrial water cycle. The consensus is strongest when the variables are related to surface temperature and hydrologic variables that depend on temperature. There is also considerable consensus for particular regions, with the strongest and most enduring projections related to high latitudes, particularly the Arctic region (see Chapters 7 and 9).

The first climate change projections were made approximately 40 years ago, led by scientists from the US Geophysical Fluid Dynamics Laboratory. These early projections were based on a low-resolution atmospheric model with very simplistic land and ocean physics. One important result was a projected significant warming at high latitudes. This is now occurring much as was projected, with considerable impacts that range from melting permafrost, earlier retreat and melting of snow, smaller snow area extent, earlier spring ice breakup on rivers, intrusion of woody vegetation into the tundra region and the death of Boreal forest trees stressed by higher temperatures. There is a documented trend in decreased snow cover extent and earlier retreat (e.g. extent in March and April) in the Northern Hemisphere based on satellite data starting in the early 1980s (see, for example [26]). The decrease in snow extent and earlier retreat of snow cover have significant potential climate impacts, because snow reflects much of the incoming solar radiation, while wet, snow-free surfaces absorb most radiation. This absorption versus reflection will alter the surface energy balance, allowing the soil temperatures to warm more, enhancing the melting of any underlying permafrost, and offering improved conditions for the growth of vegetation and expansion of shrub cover and extent (see, for example [27] and work cited therein). Understanding the impacts of high-latitude warming on other components of the water cycle is incomplete. The increased trends in river discharge from many Siberian rivers have been attributed to the melting of permafrost or the increase in precipitation that may accompany a warmer climate, but other factors have also been hypothesized.

In a similar manner, warmer winter temperatures in mid-latitudes have resulted in less precipitation falling as snow as a fraction of total precipitation – essentially having more rain days during the winter season. The latter induces sudden mid-winter runoff, causing floods and impacting water supplies that depend on spring melt. These effects have been carefully documented in the Western United States (see, for example [28]) and other parts of the globe. Finally, the retreat of glaciers, which has been the subject of many news articles, is closely linked to increased surface air temperatures.

The influence of greenhouse gases on other aspects of the hydrological cycle is more difficult to ascertain. As mentioned already, because warmer air can hold more moisture than colder air, the warmer temperatures accompanying anthropogenic climate change will result in higher humidity values, and the data indicate that this has occurred in some regions (see, for example [29]). Therefore, one expects precipitation to increase if a change in circulation does not compensate for this effect.

A number of analyses of model simulations and observations have concluded that precipitation has increased over the last century by about 2%, with significant regional variability in this change. Modeling studies have also shown that the variability of precipitation has also increased, by about 4%. With such moderate global changes, can one expect more extreme events? This seems to have occurred in tropical latitudes; although it is not clear that the total number of tropical cyclones has increased over the last 20 years, recent observations have demonstrated that there is a 30-year trend toward more frequent intense hurricanes reaching categories 4 and 5 – the two largest categories, with wind speeds stronger than 56 ms^{-1} (see, for example [30]). With regard to summertime convective (thunder) storms, it appears that their intensity has increased, resulting in more flash flooding. However, overall, the observational record regarding increased flooding is mixed, in part because of the change against a background of large variability and in part because of the short timescales and small spatial scales of many extreme rain events. Additionally, runoff involves the hydrologic process of infiltration, which varies spatially and temporally and acts to dampen the variability in precipitation through reduced evaporation.

The water budget depicted in Figure 5.1 suggests two balances: First, over long periods such as a year, for a land region, precipitation minus evapotranspiration equals river discharge from the region. The averaging over a long period assumes that the change in water storage in the region (e.g. soil moisture) is negligible. Second, also over long periods such as a year, for the overlying atmosphere, evapotranspiration minus precipitation equals the net advection (inflow) of moisture (again assuming that the change in atmospheric moisture storage is negligible). Thus, if there has been an observed increase in precipitation over a region and decreased (or no change to) evapotranspiration, it follows that river runoff and perhaps soil moisture need to increase. Both seem to be the case, at least in a number of regions. For example, researchers have reported that the flow in the Mississippi River has increased by 22% from 1949–1997 (e.g. [31]), while precipitation in the basin has increased by 10%, all while the water management activities would expect to reduce flows. River flows from northern Eurasian rivers flowing into the Arctic Ocean have also increased, but the poor hydrometeorological networks make attribution more difficult. Long-term soil moisture observations are scarce, but where available (particularly in Russia), it appears that summertime soil moisture has increased. The other major hydrologic extreme event is drought, a situation that can occur more frequently in subequatorial latitudes. Higher surface temperatures induce higher potential evaporation, and during winter and early spring, when soil moisture content is high, vegetation transpiration increases. The consequence is that, in summertime, dry soil cannot maintain its evaporation rate, resulting in lower atmospheric humidity over land and therefore less precipitation. This positive feedback process can lead to increased drought – both in intensity and frequency. Because a variety of factors can affect precipitation over these regions, including land cover change, it is difficult to attribute causes to specific drought events – for example, the drought over the Western United

States of the last few years. Nonetheless, climate change projections indicate that regions prone to drought (e.g. the Mediterranean region of Europe, Southwestern United States, portions of China) will have more frequent drought over larger areas.

Simulations of the twentieth-century hydrology over Africa also suggest that drought has increased in the Sahel (e.g. [32]), likely resulting from an increase in sea surface temperature over the Atlantic Ocean [33].

For water cycle variables that are functions of hydrologic states – namely heat (i.e. temperature) and moisture variables – the complex feedbacks described later make accurate climate change projections harder to fully quantify. For instance, evaporation depends on many meteorological factors, including temperature, moisture (humidity), and wind, whereas water use by vegetation (transpiration) includes these factors plus biophysical controls exerted by plants for growth and the availability of soil water for plant uptake. Rainfall infiltration in soil is another example: it depends on the wetness or dryness of the soil. River runoff from rain depends also on many factors as well, such as the amount of rain; the dryness of the soil before the rain event; and the topography, vegetation, and land cover of the area drained by the river. Snow is just the solid state of precipitation and frozen ground that occurs when soil moisture freezes. Yet these states affect the water cycle, with the latter preventing infiltration of melting snow or liquid precipitation, and the former allowing precipitation and water to be transferred from seasons when it falls to seasons when it melts and contributes to soil water and runoff. Thus, climate changes that affect these processes have significant impacts on the terrestrial water cycle, and the complex nonlinear process representations make accurate projections difficult. Some assessments of changes are provided in Table 5.2.

Another important difficulty is the separation of the two major anthropogenic influences, which are the impacts from greenhouse gases increases (e.g. CO_2, methane) and changes in land use due to irrigation for crops, urbanization, and deforestation.

Model consensus related to climate change

Below are some examples of model consensus from the most recent Fourth IPCC assessment (AR4) climate simulations that included a background twentieth-century climate as well as the twenty-first-century change scenarios. These models represent projections from the most advanced climate models. Shown below are global maps that indicate the number of models predicting an increase, decrease, or no change in a number of hydrologic variables, in which the difference is computed between a background twentieth-century preindustrial

Table 5.2. *Summary of trends in hydroclimatological variables over indicated periods of record, and whether the trends were regional (R) or global (land area) (G) in scale (adapted from [34])*

Variable	Period considered	Majority trend regional (R) or global (G)
Precipitation	Twentieth century	Increasing (R, G) with regional decreases (R)
Runoff	Twentieth century	Increasing (R, G)
Cloudiness	1950–2000	Little change/increasing (G)
Tropical storm (frequency and intensity)	1950–2000	Frequency – little change; intensity increases (G)
Floods	Twentieth century	Little change/increasing (R)
Droughts	Twentieth century	Little change/increasing (R)
Soil moisture	1950–2000	Increasing (R)
Pan evaporation	1950–2000	Decreasing (R)
Actual evapotranspiration	1950–2000	Increasing (R)
Growing season length	Twentieth century	Increasing (G), based on both temperature and satellite data

climate control run (designated 'PICNTRL' in Figure 5.2) and the A2 greenhouse forcing scenario (designated 'SRESA2' in Figure 5.2), which is a high-emission scenario with the projection from the period from 2070–99. A detailed description of the A2 scenario can be found in [35].

Shown below in Figure 5.2a–d are global maps that give the percentage of the models that predicted an increase (upper panels), decrease (middle panels), or no change (lower panels) for a number of hydrological variables – Figure 5.2a, annual precipitation; Figure 5.2b, annual evapotranspiration; Figure 5.2c, annual runoff volume; and Figure 5.2d, springtime (MAM) snow-water equivalence.

It is clear regarding the consensus that in the Arctic there will be more precipitation, with an increase in the spring snow pack over Eastern Siberia and decrease over Western Siberia, and an increase in runoff across the Arctic. This has implications for freshwater inflows into the Arctic Ocean and potential change in the strength of the thermohaline circulation and deep ocean water formation in the North Atlantic region (see Chapters 6 and 9). The increase in evapotranspiration in the Arctic region (Figure 5.2b) along with the increase in precipitation and runoff suggest a more vigorous hydrologic cycle in this region.

In the Mediterranean region of Europe, there is strong consensus among the models that lower precipitation and lower runoff (due to less available water), resulting in more droughts, will prevail. This pattern is also found in the Southern

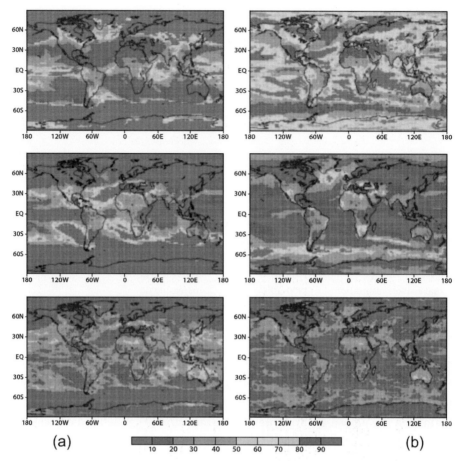

Figure 5.2a–b (a) Percentage of IPCC models that predict increased annual precipitation (top panel), decreased precipitation (middle panel), and no change (bottom panel) between PICNTRL and scenario SRES A2. (b) Same as (a), except the variable is annual evapotranspiration. (For image in color, please see Plate 5.)

and Southwestern portions of the United States and Mexico, without consensus, however, among the models regarding runoff. In fact, there is consensus of a mid-latitude drying and a corresponding wetting in the tropics and high latitudes, some of which has been borne out by observations over the last 30 years. Some research suggests that dry regions will get drier and moist regions will get wetter – suggesting that the spatial and temporal variability of the terrestrial water cycle will increase.

Finally, the decrease in springtime (March–April–May) snow pack over the Northern Hemisphere produced by the models is striking. Many of these regions depend on the snow pack for summertime water supplies and to recharge soil for crops. The decrease in precipitation in the Tibetan Plateau region has significant

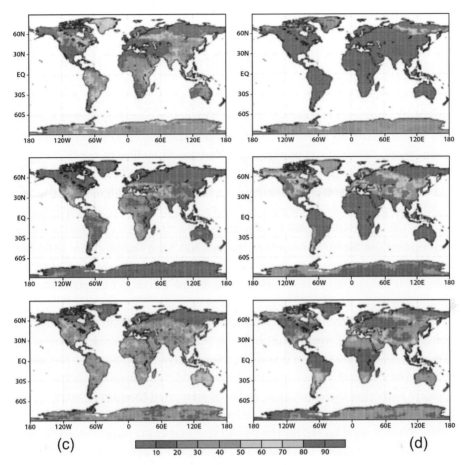

Figure 5.2c–d (c) Same as (a), except the variable is annual runoff. (d) Same as (a), except the variable is March–April–May snow water equivalence. (For image in color, please see Plate 5.)

implications for water availability in India and China and in the Western United States for water supplies for agriculture and power production.

How well can we measure and model the terrestrial water cycle?

In this section, we provide additional background on the challenges of measuring and modeling the terrestrial water cycle. Although the terms in the water budget can in theory be measured, doing so accurately remains a challenge for most of these variables (e.g. solid precipitation, evaporation, percolation to groundwater, and others) due to their variability in time and space. Modeling these variables also remains a challenge, because the processes, such as infiltration or evaporation, depend on other parameters and conditions.

Historical perspective

Interest in understanding the hydrologic cycle extends back thousands of years in China, where the earliest recorded observations were made in approximately 1200 BC. Curiosity about the water cycle in ancient Greece extends to about 300 BC, when the Greek philosopher Theophrastus (c. 372–287 BC) correctly described the atmospheric hydrologic cycle. This was extended by the Roman engineer Marcus Vitruvius at about 0 AD to include the concepts of infiltration and groundwater, derived from precipitation, and providing flow to rivers. Measurements during the Renaissance by Leonardo da Vinci (1452–1519), Bernard Palissy (1510–1589), Pierre Perrault (1608–1680), Edme Mariotte (c. 1620–1684), and others established, through field observations, that river flow comes from precipitation. Later on, Edmund Halley (1656–1742) quantified the hydrologic cycle for the region surrounding the Mediterranean Sea, and by 1800, John Dalton (1766–1844) had established the nature of evaporation and the present concepts of the hydrologic cycle, with Henry Darcy (1803–1858) filling in the missing concepts related to groundwater flow in 1856.

Routine measurements of precipitation began before 1800 in Europe and the United States and in countries like India by 1820. Such measurements were used to relate precipitation to river flow and subsequently to flood flows for the design of bridge crossings and other structures. Later, additional measurements of temperature and other meteorological variables were used to estimate water needs for water supply for growing cities and irrigation.

The observational record for the water cycle that exists today reflects the historical use of the data for flood estimation and water supply: heavy precipitation and river flows for floods and long-term average precipitation and temperature for water supply. What is well-known and can be measured quite accurately are long-term mean conditions related to the water cycle over a region: long-term evaporation is balanced by the difference between precipitation over the region and discharge from the region, usually as river flow.

Challenges in measuring and observing water-budget variables

As the region observed becomes smaller in size and the time period becomes shorter, difficulties in making accurate observations increase. Precipitation measurements at a point location suffer from the variability of rainfall over a region, and the assumption that changes in soil moisture storage can be neglected (reasonable for times longer than a year) can no longer be made. Similarly, groundwater levels and therefore water storage vary over the year. But perhaps the most difficult measurement to make at intervals useful to climate modeling is evapotranspiration. All of the above variables are influenced by the diversity of the landscape, expressed by varying topography, soils, vegetation, land use – all factors that influence the hydrologic processes described in the appendix.

It is worthwhile to discuss the difficulty with modeling and measuring surface evapotranspiration because of its unique role as a process common to the water cycle, to the energy cycle as latent heat, and to the biogeochemical cycle as plant transpiration. These cycles basically comprise Earth's climate system. The land-surface energy cycle partitions the net surface radiation into latent, sensible, and ground heat fluxes, with the small amount of stored heat usually ignored. The energy that goes into evapotranspiration (evaporation and transpiration by plants) is referred to as latent heat, and this heat gets released when the water vapor condenses. Evapotranspiration over land surfaces is determined by one of two situations: sufficient water availability with the evapotranspiration rate limited by net radiation and surface meteorological conditions, or insufficient water availability such that the soil-vegetation 'system' restricts the exchange of moisture with the atmosphere. Vegetation controls this exchange by closing its stomata. The net surface energy not used for evapotranspiration usually becomes sensible heat, while the small amount of ground heat flux, when totaled over a day, is nearly zero.

Because the net surface radiation varies over the day as the sun rises and sets and clouds come and go, the available energy for evapotranspiration varies significantly. Its variability (over time because of the radiation variability and over the landscape because soil, vegetation, and available water vary) contributes to the variability of both the water and the energy cycles. Evapotranspiration plays an important role in the feedback between the land surface and the lower atmosphere (referred to as the boundary layer). Research has shown that moist land surfaces are cooler, evaporate more, and produce more boundary layer clouds – those clouds that are about 1000 m above the ground during summer days, which can moderate land surface processes. The evaporation moistens the boundary layer, increasing its humidity, which in turn restrains further evapotranspiration. The instruments to measure accurately the evapotranspiration rate at a point have unresolved errors of about 20%–30%, and 'pan evaporation,' defined as a drop in water level from a standard pan filled with water, is an uncertain measure of the actual evaporation from natural surfaces.

Our ability to model evapotranspiration from natural surfaces shows differences of the same 20%–30% magnitude when compared with measurements or other models. The difficulties in measuring and modeling evapotranspiration over natural landscapes, together with that of properly modeling associated feedbacks with the lower atmosphere, is a serious impediment to making accurate projections of the impacts of climate change. Significant progress has been made in modeling changes in temperature due to greenhouse gases, with fairly good agreement among models. There is less agreement, however, on the changes in land surface water budget, and a significant source of the disagreement is the uncertainty related to evapotranspiration, which is poorly measured and understood. Change to evapotranspiration directly affects atmospheric water vapor, cloud formation, and, ultimately, precipitation.

Effects of land cover and its change on hydrologic processes

It is well-known that temperature and precipitation influence vegetation over land. Yet there is less recognition of the effect land cover and land cover change have on climate and climate change. Vegetation exchanges water, heat, momentum, and CO_2 with the atmosphere. These exchanges depend, to a large part, on parameters that characterize vegetal cover. The fact that 60%–70% of land precipitation (Table 5.1b) originate from evapotranspiration reveals the extent to which plants influence weather and climate.

The coupling between evapotranspiration and precipitation is accentuated in middle latitudes during the summer season, where water transpiring from plants evaporates from the surface and is recycled through precipitation over the same region or downstream. If the soil is wetter (drier) than in a normal year, evaporation and, consequently, precipitation are enhanced (reduced). This interaction helps to explain the persistence of dry or wet conditions over many months.

The situation might be reversed in the case of monsoon climates that exist in many areas just outside the tropics. The Indian and Asian monsoons are best-known, but the West African and North American monsoons are critical to the seasonal climate of the monsoon regions. In these regions, the interplay between the land and adjacent ocean plays a major role in bringing moisture and therefore rainfall into the land region. Prior to the monsoon onset, hot, dry land causes rising air (i.e. low-pressure systems) over large regions, into which humid air flows from the adjacent ocean. This large-scale flow combines with more localized convection. The mountains of these regions constrain air to rise in the atmosphere over the heated land and induce strong convective instability, condensation, and intense precipitation. The land cooling resulting from rainfall may, in some cases, suppress convection. The complexity of these climate systems – large-scale moisture flows and the influence of surface conditions before and during the monsoon – make projections of their change due to climate and land use change difficult.

The complex interactions of the wet and dry areas with the overlying atmosphere appears to influence the formation of cyclones off the West African coast that lead to tropical storms and hurricanes in the Atlantic that can later on affect the United States. An ongoing study is being led by French scientists to better understand these complex processes that lead to hurricanes and the roles of climate and land cover change [36].

Recently, much research has been performed to better understand how water management, like irrigation, affects climate. It has been observed that in specific regions (for example, over Israel or the Western United States), irrigation has changed local meteorological conditions. Over much larger areas, like Southeast Asia, where irrigation is highly developed, one can only surmise its effect on climate. Full understanding is difficult because many land cover factors are modified when irrigated crops are substituted for natural vegetation. Because detailed, large-scale climate observations are unavailable from land conditions before and after change has occurred, our understanding and climate change predictions must come from modeling studies.

Progress has to be made in understanding how large dams influence the terrestrial water storage. These large water volumes induce seepage in subsurface water and modify the water budget, but they may also change the water convergence in the atmosphere. For example, the significant decrease in Nile River flows due to the building of the Aswan Dam *must* result in a change in the long-term net advection of atmospheric water vapor over the basin because of the coupling of terrestrial and atmospheric budget.

The consequences of deforestation and replacement by short vegetation like pasture or crops is a major issue of scientific concern, particularly in semi-arid zones bordering on major deserts like the Sahel or in tropical regions like the Amazon River basin. Such land cover change involves many land-climate feedbacks. For grassland, the physiological parameters, like surface foliage and roots, are smaller than for a forest, which reduces transpiration. But in the same way that animals have pores on their skin, plants have stomata through which they exchange water vapor and carbon dioxide with the adjacent air, carbon dioxide being used for plant growth. Stomata conductance is a measure of the magnitude of these exchanges. For example, forest trees have a lower conductance (exchange) per unit area than crops do. Therefore, changing forests to cropland significantly affects the amount of water vapor released by these plants into the air. Another complexity arises from the dependence of the conductance on soil moisture deficit, which can be quite different for different vegetal covers.

Other parameters also influence the hydrological cycle when deforestation occurs: the surface albedo increases and the roughness decreases. Albedo is a measure of the fraction of solar radiation that is reflected from the surface, so albedo increase reduces the energy available for evaporation, transpiration, or surface heating. The lower roughness for grass and cropland decreases turbulence in the lower atmosphere, and, simultaneously, the friction between surface and atmosphere, thereby modifying atmospheric convergence of mass and water. Furthermore, to evaluate the effect of deforestation on terrestrial water storage, it is also necessary to take into account the infiltration rate that is larger for short vegetation. All of these mechanisms are taking place simultaneously, and to derive their net effect, it is necessary to quantify each of these feedbacks. Their importance and strength depend on the specific local climatic conditions. Research has been carried out to better understand and model vegetation and its seasonal dynamics and to couple these vegetation models with climate models from which projections related to land cover and land use change can be made. Modeling studies of the climate 6000 years ago have shown that the existence and development of vegetation can lead to a contraction of deserts by a biogeophysical feedback: vegetation enhances transpiration, which increases precipitation; this helps vegetation to grow, and vice versa from vegetation loss. Nonetheless, long-term observations of ecosystems are needed to ascertain model projection results. In the above discussion, the effect of increased albedo from changing land cover was discussed. Similarly, changing snow cover from climate change results in a lower albedo, especially in early spring, and a warmer surface as more solar energy is absorbed. This results in earlier vegetation growth, drier soils, and lower stream flows in late summer and fall due to lower springtime soil water storage from melting snow. Overall, vegetation is a central variable in understanding climate and making projections regarding climate change. In fact, studies of the influence of anthropogenic land cover changes on global climate have showed that significant temperature variations may have resulted that are similar in magnitude to those projected from increased atmospheric carbon dioxide [37].

Summary

Over the last century, changes in the terrestrial water cycle have been documented. These changes can be attributed to climate change through increased emission of greenhouse gases like CO_2 that have led to increased atmospheric temperatures and subsequently increased atmospheric water vapor, which itself is a greenhouse gas. They can also be attributed to other anthropogenic activities, such as land cover change (deforestation), draining of wetlands, irrigated agriculture, and impoundment of water by dams. These activities also affect the terrestrial water budget. The terrestrial water cycle is coupled to the atmospheric water cycle, so changes in one component affect the other. The warming of the Earth from greenhouse gas emissions will change the mean values and variability of the water budget terms, but hydrologic theory is too incomplete to project these changes with a high confidence for particular regions. Nonetheless, observations and model consensus over the next 30 to 60 years suggest increased precipitation over land, warming of the Arctic resulting in early retreat of snow covered areas and a loss of permafrost, a drying of the mid-latitude resulting in increased regional drought, and wetting of tropical and high-latitude regions. The model simulations for the next century also indicate a drying of some regions in mid-latitudes, resulting in increased regional drought and probably more frequent intense summertime rain events. Collectively, this suggests intensification of the hydrologic cycle, but with large regional variations in this change.

Recommended books and articles

T. R. Karl, R. W. Knight, D. R. Easterling and R. G. Quayle, Indices of climate change for the United States. *Bulletin of the American Meteorological Society*, **77** (1996), 279–92.

N. Arnell, *Hydrology and Global Environmental Change* (Harlow, United Kingdom: Prentice Hall, 2002).

Appendix: Major Hydrologic Processes

The major processes of the water cycle can be summarized as follows:

Evaporation and transpiration – collectively referred to as 'evapotranspiration.' Evaporation is the change of water phase from liquid or solid form to vapor, while transpiration is the water usage by plants through extraction of soil water by roots. Sources of free water available for evaporation include soil water; water from lakes, rivers, and wetlands; and water on vegetation left from recent precipitation events. Evapotranspiration results in drying the soil column and humidifying the atmosphere. In wet conditions, evapotranspiration is limited by incoming solar radiation and meteorological conditions, while in dry conditions, evapotranspiration is limited by the soil and vegetation resisting and limiting water loss. This limiting condition makes the measurement and prediction of evapotranspiration challenging. Evapotranspiration requires energy, obtained from solar energy, which is later released as heat when water vapor condenses.

Condensation and precipitation. Condensation is the change of water phase from vapor to liquid (rainfall, dew) or solid (snow, frost), which occurs when the air mass containing the water vapor becomes saturated. Saturation occurs when the air mass cools sufficiently, such as air rising over a mountain, since warm air can hold more water vapor than cold air, or when an air mass is compressed (converges) into a low-pressure region of the atmosphere. Precipitation is the fall of condensed water in liquid form (rainfall) or solid form (snow). Condensation of water vapor leads to the release of the energy that originally was used for evaporation and is referred to as a 'release of latent heat.' Most condensation and subsequent precipitation are due to the convergence of air masses and result in rainfall that is highly variable within a region and during a rain event. Precipitation that falls as snow requires energy to melt, so it is an absorber of energy, and the subsequent liquid water will either infiltrate or run off, as described below.

Infiltration. Precipitation falling on the ground divides into a portion that enters the soil and a portion that runs over the ground to gullies, streams, and eventually into rivers. The hydrologic process that controls that portion of water that enters the soil is known as infiltration. The rate of infiltrating water depends on the rain rate, the antecedent soil moisture, and the soil characteristics, like the fraction of sand and clay, its texture, and so forth.

Percolation. Percolation is the process that describes the movement of soil water from the near-surface zone that is unsaturated (i.e. water and air coexist in the voids existing between soil particles) to the saturated (or groundwater zone), where the voids are filled with water.

Groundwater. Groundwater (or the saturated zone) refers to the portion of the soil column, where the void between soil particles is completely filled with water. This is contrasted with the unsaturated zone, where the voids are filled with both water and air. The 'groundwater table' is that depth where the water pressure within the soil (due to gravity forces) equals atmospheric pressure. Deeper in the saturated zone, the pressure is greater than atmospheric pressure and closer to the surface, and in the unsaturated zone, the pressure is less than atmospheric pressure.

Runoff and river discharge. Runoff is usually referred to as that portion of a precipitation event that immediately runs across the ground surface, down gullies, and into streams. Stream and river flow during non-rain periods is augmented by water, previously infiltrated into the soil column, and percolated into the groundwater zone, draining from the soil column into streams and rivers.

6

Ocean and climate

CARL WUNSCH AND JEAN-FRANÇOIS MINSTER

Every man, wherever he goes, is encompassed by a cloud of comforting convictions which move with him like flies on a summer day.
— Bertrand Russell

Ô mer, nul ne connaît tes richesses intimes...
— Charles Baudelaire

Introduction

The ocean contains almost all of the water on the planet; is a vast storehouse of heat, carbon, and biological nutrients; moves these fields around in complex ways; and exchanges them with the atmosphere and land. Understanding how this immense fluid/biogeochemical system interacts with the rest of the climate system is a very challenging problem in both observation and theory. A full understanding of how the ocean works as a fluid biogeochemical system and the nature of its variability is just beginning to emerge. This knowledge is

Carl Wunsch is Cecil and Ida Green Professor of Physical Oceanography at the Massachusetts Institute of Technology in Cambridge, MA, where he has been a faculty member since 1967. His interests are the general circulation of the ocean, including both observations and theory, and the oceanic interaction with the wider climate system. He has been involved in the development and use of satellite altimetry, acoustic tomography, inverse methods, and in the World Ocean Circulation Experiment.

Jean-François Minster has been mainly working on biogeochemistry of the ocean (trace metals, carbon cycle) and on satellite altimetry. He directed the space oceanography and geophysics laboratory in Toulouse from 1985 to 1996, then the Institut National des Sciences de l'Univers of CNRS from 1996 to 2000, and was CEO of Ifremer from 2000 to 2005. He was appointed as Science Director of Total in 2006.

critical for the study of climate variability, the detection of long-term changes, and the possible eventual prediction of how it might shift in the future. Here we discuss how the rich new information about the ocean obtained in the last decade influences the debate about what the future holds.

From the point of view of anticipating the consequences of global change, the ocean is plainly an extremely important element. The ways it changes in the future are likely to be a major determinant of the Earth's climate as a whole. In this essay, we describe some of what is known about the system, outline the complexity of the problem of describing the ocean as it exists today, and forecast how it might be different in the future.

Until very recently, observations of the ocean were so difficult to obtain and the numerical modeling problem so formidable that it was commonly rendered as a grossly oversimplified system. The oversimplification in turn invited oversimplified discussion of how it would change under greenhouse gas effects. Similarly, reconstructing the ocean state in the geological past, which contains powerful clues as to the natural variability of the climate system, also led to schematic, rather than realistic, explanations of oceanic physics, chemistry, and biology. Today, advances in both observations and ocean modeling permit a revised view of the ocean as a complex, turbulent fluid in which the degree of true predictability remains unknown (some systems, chaotic ones like the weather, cannot be predicted for indefinite periods). Much more is now known about the problem of making useful forecasts of future oceanic states, and we can begin to grapple with the observational and theoretical requirements that will be necessary to reduce our uncertainties and to formulate public policies that are realistic in terms of what the science can say today.

What is clear, however, is that the ocean is capable of important influences on climate, and the uncertainties about what it will do in the future have to be an important element in societal decisions. One cannot simply wait until the evidence for major oceanic changes is overwhelming – by then it will likely be too late for a useful response.

The historical perspective

Observational background

Understanding the ocean requires observing it and describing its climatological state. One can usefully divide the problem of observing the ocean into three interconnected sub-problems of determining: 1. the surface properties and the exchanges there of momentum and energy with the atmosphere; 2. chemical and physical properties (temperature, salinity, oxygen, CO_2 concentrations, etc.)

in the oceanic interior; and 3. the movement (flow and mixing) of all these properties. Surface properties of the ocean have been accessible to a large degree since antiquity, and with the advent of artificial satellites, many useful parameters (e.g. temperature, roughness) have become routinely determinable from space.

The major issues lie with problems number two and three – properties and flow below the sea surface – and the great difficulties encountered have in large measure dictated the inferences made about how the ocean 'works.' Oceanic surface properties, which provide the most immediate coupling with the atmosphere and cryosphere, are in large measure controlled by subsurface motions; thus, solving problem number one does not provide an adequate description of the ocean. As we will see, some of the inferences made about the three-dimensional ocean have been dictated primarily by what was measurable or inferable rather than from complete understanding.

The central issue for anyone attempting to understand oceanic properties below the sea surface is the fluid opacity to electromagnetic radiation: one can neither send radio waves through the ocean nor peer very far into it from the sea surface. Most properties of the ocean can be measured only by placing an instrument at specific locations and depths, making the observation, and returning the measured data by some means to an observer at the surface or ashore. One cannot radio measurements back to the surface, and until extremely recently, all observations required sending a ship to any location of desired data, lowering an instrument physically to the requisite depth, and either sending the data up a wire to the ship or recording them internally in the instrument until they were returned to the surface.

Observing a global ocean using ships is extremely expensive and slow: a typical crossing of the North Atlantic (a small ocean) with a modern oceanographic ship stopping to sample typically takes a month at a cost of approximately $50 000 per day. Such measurements have been, and remain, comparatively rare. The first global reconnaissance of oceanic properties took place over many decades, involved many different countries, and was quite limited in scope. Early in modern oceanic exploration (commonly dated as beginning at about 1870), it was recognized that oceanic interior properties, such as temperature and salinity, vary comparatively slowly over vast distances and depth and with only small changes over long times (see Figure 6.1). This feature meant that observations scattered over many decades could be combined to produce a large-scale depiction of the apparent time-average ocean. Temporal changes in particular could not be addressed and appeared to be unimportant. Similarly, because observations extending to great depth were expensive and often difficult to obtain, many seagoing expeditions stopped their observations well above a depth of 1000 m, on the more or less explicit supposition that little or nothing

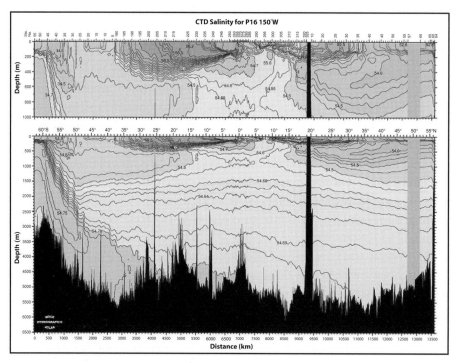

Figure 6.1 Salinity section (from the World Ocean Circulation Experiment – WOCE Atlas series) called P16 down the central Pacific Ocean. The salinity changes from about 36 (near the surface in the tropics) to about 34 (the large light domain at depth). A salinity of 34 corresponds to a density contribution of about 3.4%. Note both the large-scale property distributions and the presence of smaller-scale fluctuations, which in some cases are purely 'noise' but in other cases represent, through their spatial derivatives, important density and hence flow fluctuations. The large-scale structures can be depicted using data scattered over many decades, but the smaller scales change rapidly with time.

interesting was occurring below that depth (the ocean has an average depth of about 3800 m).

Obtaining observations of oceanic biological and chemical parameters is still more difficult than for the physical ones. Although the measurement of nutrients and dissolved oxygen has been nearly routine since the 1950s, measurement of other parameters, such as the ionic forms of dissolved carbon and other tracers, could only be done with the necessary accuracy after the 1970s. As a consequence, the first attempts at a global description occurred only in the 1970s (the Geosecs program). Because of the need to take large water samples under very specific conditions (adapted to each tracer), it was based on a series of only 422 stations for the whole ocean. Thus, the ocean description that was derived from these data had to rely on the hypothesis that the ocean was steady, with no significant small-scale features.

Theoretical background

Property fields do not show how the ocean is moving. Determining how it moves can be done only by using the equations of fluid dynamics for a rapidly rotating spherical shell (the Navier-Stokes equations on a sphere), supplemented by those of the relevant thermodynamics and interacting with the overlying atmosphere. The complex lateral boundaries of the ocean must be accounted for. Interaction with the atmosphere takes the form of exchanges of momentum (wind-stress driving), heat (enthalpy), freshwater, and gases such as CO_2. Solving the resulting equations (their form can be seen in any oceanographic textbook; see e.g. Open University, 2001, for a simplified version) is one of the most formidable problems in mathematics, and no general solution has ever been found. Instead, fluid dynamicists and physical oceanographers look to simplify the equations so that only the most important effects for the problem at hand are retained. Thus, for example, in studying the ocean circulation, one would normally omit mathematical expressions of surface tension (important for describing ripples) or compressibility (leading to acoustic waves) from the equations being used (both of these phenomena are present in the full equation set).

The inference from observations that the ocean moved only on very large scales and did not fluctuate significantly in time provided an immense simplification to theoreticians attempting to understand it – they could omit from their equations the representation of time changes and of 'small' scales. Although 'small' was not clearly defined, it might be understood as anything varying significantly over distances less than about 1000 km. The resulting simplified equations of motion could be solved in many interesting and useful cases and provided a sophisticated and often counterintuitive picture of the ocean circulation as a basically steady, large-scale flow driven primarily by the winds, and heating and cooling at the surface. In many ways, the theories produced property distribution and flow structures appealingly like what the observations appeared to give. By about 1970, but further refined over the following years, an apparent zero-order understanding of the ocean emerged. Figure 6.2 [38] is representative of the type of steady flows in deep waters depicted by theory.

The climate consequences of this oceanic circulation were not very much explored, in part because climate had yet to become the major scientific focus that it eventually did beginning in the 1980s. Oceanographers had concluded that the ocean circulation contributed only marginally to the global heat balance, and there was little or no interest or understanding of the global water cycle or the exchange by the ocean of greenhouse gases such as carbon dioxide. The ocean

Figure 6.2 The solution obtained by Stommel and Arons in 1960 [36] for the abyssal flow of the ocean. Note that it is large scale and steady and suggests the flow in the deep water is predominantly meridional and slow. One might compare it to Figure 6.4, which shows the estimate from observations of the actual flow across the North Atlantic Ocean.

was widely regarded, particularly by meteorologists, as primarily a static reservoir of heat and moisture (a 'swamp') with little or no dynamical implications for climate control or change.

Breakdown of the large-scale steady picture

The comparatively elegant and simple picture of the ocean circulation began to break down in the middle 1970s, but it would be incorrect to assert that the older picture has been completely replaced. Some of the classical picture retains validity in describing basic underlying principles, and the oceanographic community has been slow to embrace the conclusion that much of the classical picture is incomplete – almost to the point of invalidity.

The classical picture – of a slowly changing, laminar ocean – began to dissolve when electronic technology developed sufficiently far, in the middle 1970s, to permit the first true time series of oceanic properties to be made at arbitrary locations and depth. The technical requirements were for instruments capable of *in situ* storage of measurements over months to years and reliable methods for recovering moorings left behind by a visiting ship. It slowly became obvious that essentially all elements of the ocean circulation – temperature, salinity, and

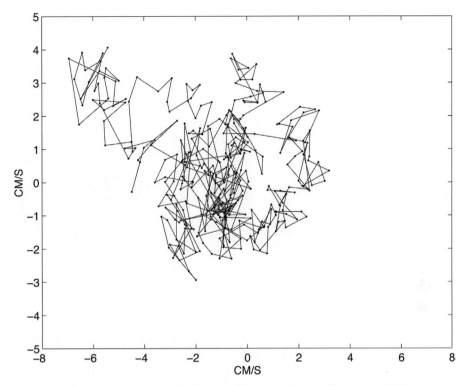

Figure 6.3 Trajectory of a fictitious particle having the velocity from a 24-hour average for each day over one year. The record was obtained at 1528-m depth at a position 27.3 N, 40.8 W in the North Atlantic as part of the Polymode program. The result is typical of open ocean records (although this one is quiet). This particular record does produce a mean displacement toward the northwest. If the high-frequency components (periods shorter than one day) are included (not shown), the figure is an extremely dense cloud.

flow fields – were undergoing continual and rapid temporal changes, and that these changes appeared to decorrelate (become essentially independent) when separated by no more than about 50 km in the horizontal. As observations slowly accumulated over the next decade, it emerged that the ocean is best regarded as fundamentally turbulent and in fact time-varying on all timescales and spatial scales. The idea of 'turbulence' is perhaps best supported by the inference that fluid motions varying on timescales of over 100 days and on spatial scales of 100 km have more than 100 times the kinetic energy (10 times the speeds) of the large-scale time-mean flow. Open ocean current meter records are dominated by seemingly random motions in which the time average is rarely statistically significant (Figure 6.3). Some of the spatial complexity can be seen in Figure 6.4 [39], which displays an estimate of the absolute north-south component

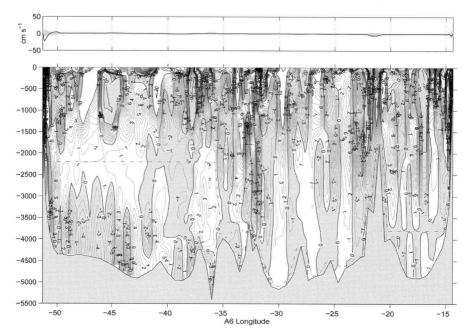

Figure 6.4 After Ganachaud [39] showing an estimated near-snapshot of the flow field across the North Atlantic Ocean at 24 N. Stippled areas are flowing southward, clear areas to the north. Thus the flow changes direction both with longitude and depth. Obtaining a stable, statistically significant mean mass flux is very difficult in the presence of such energetic structures, many of which are time dependent. Determining the heat and other scalar properties carried by this flow involves the integral of the product of this field with temperature or other tracers.

of flow across the North Atlantic Ocean. Understanding the ocean circulation, and possibly ultimately predicting it, depends on understanding these flow features, why they are present, how long they persist, and how much of the properties we care about are carried by them. Ocean 'sampling' is clearly a serious problem both in space and in time.

Recognition that the ocean was filled with an eddy-like motion, superimposed on the larger scale, quasi-steady fields that had been the focus of attention for 100 years, raised a number of difficult questions. One might wonder how such a phenomenon could have escaped the attention of the clever oceanographers who had worked prior to about 1975. Some oceanographers clearly did recognize the presence of strong time variations (e.g. B. Helland-Hansen and F. Nansen before 1920), but they lacked the technology capable of dealing with it, and thus the wider oceanographic community came to assume that it was unimportant.

In addition, there appear to be two major reasons for this oversight. First, the spatial separation of observations tended to be much greater than 100 km horizontally – they were time-consuming, expensive, and difficult to make, and the large-scale patterns were still visible with sparse spatial and temporal sampling. Second, the equations of motion show that in general, the movement of fluid depends not on the distribution of temperature and salinity but on their derivatives (that is, their time and space changes). The human eye is a wonderful instrument for seeing large-scale patterns in the presence of noise – evidently, there is an evolutionary advantage of such pattern recognition (e.g. [40]). The eye is not very good at quantifying the values of small-scale fluctuations whose differences tend to dominate the fluid motions. Yet it is now clear from the study of turbulent fluids that small-scale motions can affect the large-scale patterns and property transports. In other words, the small-scale transports, such as Figure 6.4, do affect the property distribution, such as observed in Figure 6.1, in addition to the large-scale transport picture of Figure 6.2.

In addition to the inference that the oceanic kinetic energy was dominated by flows varying over 100 km, it also gradually appeared that even the large-scale patterns of temperature, salinity, and various chemical properties as well as the superimposed large-scale flow fields were themselves changing with time, albeit on timescales too long for easy, direct human perception. An exception to the long timescale problem is El Niño, in which large fluctuations occupy the tropical Pacific Ocean with very direct coupling to atmospheric changes of immense importance (floods, droughts, heat waves, etc.) occur at intervals of three to nine years. Some of the difficulty of this subject in general can be appreciated by recognizing that El Niño was known to the Spanish conquerors of South America as early as the middle sixteenth century [41]. It took another 400 years, until the late twentieth century, before it was understood that El Niño was a near-global phenomenon of gigantic economic and social importance. El Niño is not at all subtle, and it makes one ask what less dramatic fluctuations are also taking place without human notice? Addressing that question of detecting and understanding the various possible processes at play in the large-scale climate variability on various timescales is the present major issue of ocean climate studies.

Toward an ocean-observing system

If one is to observe climate change, an observing system is required, one that is designed to obtain adequate samples. At the end of the 1970s, the World Meteorological Organization (WMO), in collaboration with the oceanographic community, launched a series of programs that led both to much-improved estimates of the ocean transport and to the present-day ocean observing system. Although a major goal was improvement of ordinary weather forecasts, a second objective was to understand El Niño and its impacts. This objective led to the

TOGA (Tropical Ocean and Global Atmosphere) program, which ran for 10 years, beginning in 1985. The legacy of TOGA is not only a much more detailed understanding of El Niño dynamics; it is also a system to detect the phenomenon early and predict its evolution. This system is based on a series of moorings in the tropical Pacific Ocean (and also a few others in the tropical Atlantic). These moorings transmit 'real time' observations of a variety of parameters in the atmosphere and in the first 200 m of the ocean. They are complemented by *in situ* measurements on board commercial research vessels and island stations. The observing system also depends heavily on satellite measurements, including those of wind stress, sea surface temperature, surface chlorophyll, and the elevation of the ocean surface. The system includes numerical models representing the ocean, which are made realistic by their adjustment to the observations. During the last 10 years, this system proved successful in detecting the appearance of El Niño conditions quite early (approximately six months in advance) and in making reliable predictions of its evolution 6 to 12 months ahead.

The third objective of the meteorologists and oceanographers was more specifically about climate. The need for a far better understanding of how the ocean interacted with the climate system as a whole led to the World Ocean Circulation Experiment (WOCE). It was designed to provide a comprehensive near-snapshot of ocean observations over about five years. Among the large variety of *in situ* observations obtained during WOCE were the hydrographic (temperature and salinity) sections spanning the oceans from the comparatively short interval of 1989–97. These observations depict all of the oceans approximately every 10°–20° in latitude and longitude. They allowed estimation of ocean transports during the WOCE period with unprecedented reliability. The water, heat, salt, nutrient, and oxygen transports within the various water masses are now known with a precision of the order of 20%; previously the uncertainty often approached 100%.

Finally, these sections are a reference field permitting comparisons with ocean transports estimated, in some places, at other periods, both before WOCE (as far back as the 1930s, albeit crudely) and in the future. It is indeed found that the heat storage and transport of all other properties vary with time. However, because of the very limited data sets, it is impossible to construct a global picture of such changes and thus decide whether they represent trends or natural variability of the ocean.

Along with the implementation of WOCE and TOGA, oceanographers systematized the use of satellite data. In particular, precise altimetry data became widely available with the launch of TOPEX/POSEIDON in 1992 and Jason-1 in 2001. These satellites became a unique tool for observing variations of ocean currents on all space and timescales including the time-average. Variations on the basin scale are indeed observed, thanks to this 13-year altimetric data set. Many of these variations are explained by changes of the wind-stress pattern and adjustment of the ocean through various dynamical mechanisms. The specific case of heat storage and mean sea level will be described later.

WOCE has also helped accelerate two other major developments: one in numerical modeling of the ocean circulation to be discussed below, and the technologies for *in situ* observation.

The most successful new tool for widespread *in situ* observation that emerged in the recent years is the profiler. This autonomous system oscillates between the deep ocean (at about 2000-m depth) and the surface at approximately 10-day intervals. While at the surface, it transmits its position and *in situ* temperatures and salinities through satellites. The flexibility of this tool is such that more than 2000 such profilers are at present deployed by many organizations from around the world in what is known as the Argo experiment. Efforts are being made to increase float numbers to 3000 by the end of 2007. Data are assembled and processed in near real-time and made publicly available. For the first time, a global, repeated, three-dimensional set of hydrographic observations exists. Of course, such a relatively limited number of profilers does not allow separation of the turbulence field from the large-scale time-average field. Indeed, the typical distance between two profilers is in the order of 300–500 km, which is larger than the scale of the eddy field, as observed in Figure 6.4. The separation of the fields is accomplished by merging these and other data, particularly altimetric observations from the several satellites now flying, with the equations of motion describing the fluid flow. A longer-term problem is the design and sustenance of a fully adequate global ocean observing system (see Chapter 12).

The paleo-oceanographic picture

Study of the climates of the past is an important and interesting source of information concerning how the ocean has previously been different from what we observe today, how rapidly it can change, and how it might appear in the future. The study of geology has led to the inference that Earth's climate was very different in the past. With the availability in the last 30 years of cores drilled into the seafloor and the Greenland and Antarctic ice caps, the magnitudes and timescales of inferred changes became much clearer. For example, it is now known that over the last approximately 800 000 years, there were seven full cycles of glaciation/deglaciation of the high-latitude Northern and Southern Hemispheres. In central Greenland at least, it has also been inferred that abrupt warmings of the order of 5 °C (9 °F) had occurred on timescales of perhaps as few as 10 years.

A very large effort has been directed toward understanding how such changes could occur – and much of the discussion has focused on the ocean. The ocean came to prominence because sediment cores do reveal changes of the ocean state (water temperature, salinity, primary productivity) that are apparently associated with changes of the volume of water in the main polar ice caps. Second, it was difficult to envision such rapid changes being controlled from land-based areas. Thus, the irony was that the quasi-geological picture of a sluggish and slowly changing ocean constructed by physical oceanographers was turned on its head by paleoclimatologists and geologists, who began to invoke a very reactive ocean as a rationalization for much of what they inferred from ice and seafloor cores.

The particular oceanic scenario that is widely employed to explain much of climate change on timescales ranging from decades to hundreds of thousands of years and beyond has been the result of an extreme simplification of the already oversimplified quasi-steady laminar model into a concept called the 'global conveyor' [42]. This notion has been widely reproduced in the public press and even in two successful Hollywood movies. It is based on the assumption that the ocean circulation can be understood as a large-scale 'conveyor belt.' In this model, cold water sinks in the high-latitude North Atlantic, flows at great depth into the Southern Ocean, and on into the North Pacific, where it upwells. The fluid is then supposed to return from the Pacific into the Indian Ocean via the Indonesian passages, around the Cape of Good Hope, and on into the North Atlantic, thus completing the circuit. Rationalization of climate change is effected by postulating fluctuations, including even 'shutdown' of the system, at least in the North Atlantic part. Such weakening or shutdown is supposed to be effected, e.g. most commonly by the flooding of freshwater into the northern North Atlantic (traces of such floods are recorded in the land geology of North America).

The attraction of such a scenario is apparent: it appeals to the common intuition that the ocean circulation is driven by high-latitude convection (the sinking of dense water) and the accompanying transport of heat into the North Atlantic, which then controls the global climate system. Furthermore, the concept greatly simplifies the problem of observing the ocean – all one would need to do would be to measure the intensity of the 'conveyor' at one location (usually the comparatively accessible North Atlantic), and in an essentially one-dimensional flow, all global consequences immediately follow. This scenario also appears to eliminate the need for complex calculations with the fluid-dynamical/thermodynamical equations – insofar as its structure and behavior can be inferred by simple processes.

Use of this picture as a realistic depiction of the ocean circulation is the subject of debate (e.g. compare to Figure 6.4). Yet beyond the purely descriptive aspects, one can understand its consequences in the light of the theoretical and observational studies already described.

The ocean does conspire with the atmosphere to transport heat from the tropics toward the poles (Figure 2.2), where it is radiated back to space (the combination might be thought of as the true global conveyor). Predicting how such a system would change if, e.g. one were to add large quantities of glacial meltwater at the upper edges, requires a complex set of calculations. A number of such numerical simulations of the coupled ocean-atmosphere system have been made with this intention. They do reproduce the shutdown of the North Atlantic parts of the circulation when freshwater is added to the Arctic Ocean. However, it must be recognized that the models are not yet realistic enough to be taken as definitive answers on this issue. For example, they do not have enough resolution to represent ocean turbulence phenomena, which are so critical in the explanation of the Gulf Stream intensity, the deepwater formation processes, or the special dynamics of the upper 100 m of the ocean. A publication bias also exists – results from other models that do not exhibit such behavior are deemed 'uninteresting' and thus fail to be noticed even when they are published.

Second, one of the strongest inferences about the circulation from theory (and altimetric observations) is that the wind field is the primary controller of the ocean circulation. All of these theories support the conclusion that if one sought to change the ocean circulation rapidly and efficiently, one would do so by shifting the wind field. Changing the heat and freshwater exchange between the ocean and atmosphere would surely also affect the ocean circulation, albeit in a less-direct and -immediate manner. Yet, if one postulates such a change having taken place that leads to a significant climate shift, one would expect the atmosphere, and its associated surface wind field, to also change in important ways. One cannot ignore the wind field change in determining what the ocean would actually do and what the ultimate climate state would look like.

The practical impossibility of deducing the ocean system behavior by pure thought shows that one should be careful in producing definitive assertions about how the ocean will affect climate as various external parameters are varied until such time when models are far more sophisticated than they are today.

Numerical models

Because the ocean is a complex fluid and chemical system, one must rely on computer simulations to fully describe it. Predictions can only be made with the help of such models. Numerical models of the ocean, commonly called general circulation models (GCMs), have been employed since almost the earliest days of computers following World War II. Over the past few decades, they have grown immensely in skill and realism to the point where they have many practical uses, including the ability to aid in the understanding of climate-related processes.

Their use on climate timescales in particular does, however, raise serious questions. A numerical model of the ocean or the climate system as a whole is a rule for calculating the state of the ocean at some short time in the future (typically one hour or less) given the state today and the forces acting on the ocean. One then steps forward in time, hour by hour, until the calculation is carried as far into the future as one wishes or can afford in computer costs.

Carrying out such calculations requires specifying the oceanic state, which includes three components of velocity: the temperature, salinity, and pressure. Prescribed external forcing includes winds, heating, cooling, evaporation, precipitation, and runoff. If the entire climate system were to be modeled, the state would expand greatly to include, e.g. all the atmospheric and cryospheric parameters as well as those pertaining to land, such as the biosphere and groundwater. Computer codes can calculate a model version of the climate system as it might be, arbitrarily far into the future. How accurate will such a calculation be? It is difficult to say, as a number of sources of error are easily recognized and include errors in the state at the beginning of the calculation, errors in the forcing terms, approximations made in converting the underlying partial differential equations to numerical form on a computer, and others too technical to describe here.

Because numerical models tend to accumulate errors from various sources, there is a gradual loss of accuracy with time. A good model run for one year will, on average, have a greater accuracy in its various elements than one run for 100 or 1000 years. The issue for ocean and climate models generally is that there is little understanding of which elements of the models can be relied on after an arbitrary elapsed time. Suppose one defines an element of interest (it might be oceanic heat content or, equivalently, the mean oceanic temperature, H). One needs to know H with an accuracy of $\times\%$ and one needs to know at what future time, T, the error criterion will be violated. Can a model calculation of H be relied on to this accuracy after T = 1 year, 10 years, or more? Almost nothing is known about error growth, except that T is not infinite.

Today, calculations are made of the ocean circulation, either in isolation or when coupled with atmosphere and ice models, and integrated out to 1000 years and beyond. Sometimes the behavior of such models is very interesting, involving, for instance, complete reversals of elements of the modern circulation and/or bifurcations and other phenomena. Unfortunately, however, it is difficult to know which elements of these models can be regarded as physically well understood and which might only be artifacts of oversimplified physics, reduced dimensions, or accumulated errors.

To cope with this problem, ocean and climate modelers use several approaches. The first approach is the comparison of model simulations with observations of the last decades; water mass and frequency of ENSO events are typical tests. Each numerical model shows its own qualities and deficiencies in such comparisons. These tests are necessary for placing confidence in model projections for the next several decades, but they are not sufficient. A set of simulations can also be made with a given model, starting from a set of initial conditions that are varied to describe their uncertainty. Sometimes simulations are made with a set of different models in which the description of the processes is simplified in different ways. This major effort provides estimated ranges of predictions, permitting estimates of their robustness; these approaches are the basis of most climate predictions, but such multiple model runs are limited to comparatively short time intervals because of the very high computational costs. Another difficulty is that most models employ similar simplifications: in particular, as mentioned before, they tend to be too crude to represent the short-distance scale mixing and energy dissipation phenomena that are so critical to the intensity of major currents, heat or carbon transport, or deepwater formation. Thus, the reliability of ocean-alone or coupled ocean plus atmosphere models is still poorly understood.

Other observed changes and sea level rise

Because the ocean is the major reservoir of freshwater and one of the dominant carbon and biodiversity reservoirs on the planet, any change in it has potentially important consequences for many aspects of the Earth system. Beyond the case for water masses and ocean transports, one observes a number of changes in the ocean on the decadal timescale. One is the remarkable reduction of the Arctic sea ice coverage: satellite observations show a continuous decrease of its surface during the last 30 years of about 8% per decade in the summer season. One also observes changes in the geochemistry of the ocean and its ecosystem, from plankton species to large carnivores that occur in large areas of the ocean. Any of these changes raises questions of observability and sampling, of understanding of the processes at play, and of separation of natural variability from man-induced trends. As an example of the issues that arise, let us briefly examine the problem of sea level change, because the economic and political consequences of a rise along the shoreline are remarkably large (e.g. [43]).

Global mean sea level rose by about 130 m in the interval between the last glacial maximum at about 20 000 years ago and today as a consequence of the melting of the great Northern Hemisphere continental ice sheets. As the world warms, one expects further melting of mountain glaciers and possibly of portions of the Greenland and Antarctic ice sheets. To the extent that the ocean itself warms, the expansion of the water column contributes to a further rise in sea level. (Note that expansion by warming of the global ocean since the glacial maximum

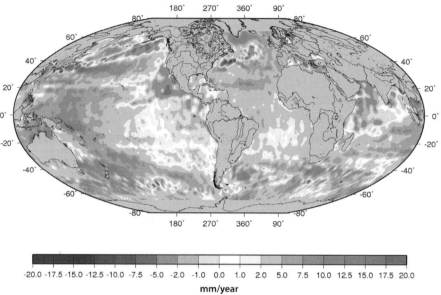

Figure 6.5 Estimated trend in sea level (mm/y) from altimetric satellite measurements for the time interval from 1993–2005. White regions are areas of no data (after A. Cazenave and S. Nerem [43]). Note the very complex spatial pattern and that sea level actually falls in many regions. (For image in color, please see Plate 6.)

induced by the associated climate warming has been only a small fraction of the contribution from the melting ice.)

How is sea level behaving today? Answering this question requires, first, an understanding how one observes it. The distribution of tide gauges with records exceeding 20 years is heavily skewed toward Northern Hemisphere continental coastlines. If global sea level is determined on average from these instruments (a complex process involving, e.g. corrections for vertical movement of land; see [44]), it results in about 1.8 mm per year from about 1900–1990. Such a number can be quite reliable if the patterns of sea level change are large enough in scale that the nonuniform sampling distribution does not bias the result. What is the pattern of sea level change?

Altimetric satellites that became available in 1992 provide a data set that is coming close to global uniform coverage. The global mean from 1992–2005 is about 3.1 ± 0.4 mm per year. Figure 6.5 shows the estimate of the trends in sea level as a function of position made by Cazenave and Nerem [45]. The striking feature of Figure 6.5 is the remarkably complex pattern in which sea level is actually falling over large areas. The issue of whether the tide gauge estimate can produce a useful global average for times prior to the altimetric satellites is clearly a very serious one. The immediate impact of sea level change is evidently quite different in regions of strong rise, such as the northern North Atlantic and the western Pacific, than it is in regions of significant fall, such as the northwestern Indian Ocean. Whether these patterns will persist indefinitely is not known. Figure 6.5 also raises two questions: What is the cause of the global sea level rise – melting ice, ocean warming, or some combination? What controls the complex spatial distribution of sea level change?

Answering these seemingly simple questions is at present extremely difficult. Determining ocean warming would appear to be reasonably straightforward – one simply measures the temperature everywhere at two different times, averages the results, and takes the difference. The main difficulty is that the ocean is very large and deep and past data are very inhomogeneously distributed. Scientists debate whether adequate data exist from past decades to sufficiently accurately calculate the average temperature and salinity in the last decades to detect the changes. Setting aside such troubling questions, analysis of the height change deduced from gridded ocean temperature data sets over the period of altimetric data seems to show a strong correlation with the pattern of Figure 6.5. Some estimates suggest that about 60% of the mean sea level signal in that recent period is explained by heat storage changes, but other estimates are as low as only 10% of the total. A significant, and conceivably dominant, fraction is caused instead by the addition of and redistribution of freshwater from melting glaciers. At least one analysis of temperature data over the period from 1950–98 showed variations of the mean height change due to warming of only 0.1–0.2 mm per year, implying that the rest was coming from melting ice. Conflicting results are an indication of the complexity of the problem of piecing together the partitioning of sea level change between warming and ice melting and makes it difficult to predict what the future holds.

The future

The ocean is important in the climate system because of its storage of heat, freshwater, carbon, etc. We know that it is changing in complicated ways. Only recently (after about 1992), with the advent of observation systems able to define the global system, has there been any capability to depict the circulation and its fluctuations, and we now understand how intricate it really is. Unlike the rather simple picture that still exists in most textbooks, the ocean has been shown to be a much more exciting fluid flow capable of storing chemical, biological, and physical properties and moving them around over long distances and times. The challenge to even the most elaborate numerical models is a very great one.

We know that the ocean is changing and can guarantee that it will continue to change in the future. Its changes have immediate impacts (such as ongoing mean sea level rise and reduction of the Arctic sea ice coverage) and potential future impacts, such as a shift in the rate of uptake of carbon, on the frequency of extreme storm events, and on marine ecosystems. Some of the observed changes are likely a consequence of anthropogenic effects, and they may well become more prominent in the future. The possibility of catastrophic shifts cannot be ruled out, but they should not be given an unwarranted weight. Slow, steady, undramatic shifts can, when maintained for decades, lead to societal impacts that are very large and more probable than truly catastrophic jumps (e.g. sea level rise or small changes in the uptake of carbon dioxide). Mankind is essentially living on the shores of the ocean, and our economies are strongly dependent on climate and also on ocean ecosystems and resources. Thus, we cannot ignore ocean changes and must have the capacity to anticipate how the ocean might change further.

Any discussion of public policy must take into account the possibility that the ocean today is undergoing transitions to different states or is nearing thresholds where it might change in important ways. At the same time, the complexity of the fluid ocean is so great that it is very difficult for scientists to make definitive statements about what has been happening in the past or what is happening today that are sufficient to predict what the ocean will be like in the future. The complexity of the ocean means that great uncertainty about global change will persist for many years. Will sea level rise diminish or accelerate? We are not in a position to say with any great confidence what is to happen, but we can make informed statements about the most likely possibilities and estimate their societal consequences. Decision making in the presence of serious uncertainty is neither novel nor intractable.

What is needed so that future generations will be able to make better-informed decisions? We must maintain and extend our existing observation systems, particularly precise altimetric satellites and *in situ* observing systems similar to the profilers – we must enable our successors to know how the ocean has changed so they can begin to understand these changes. Similarly, we need more aggressive development and systematic testing of ocean and coupled climate models so that their skills are known and quantifiable. For this to be achieved, adequate computer and manpower resources will have to be provided. More generally, there needs to be wider appreciation of the challenge of understanding the ocean and the importance of ultimately doing so.

Recommended books

S. G. Philander, *Is the Temperature Rising? The Uncertain Science of Global Warming* (Princeton, NJ: Princeton University Press, 1998).

Open University Course Team, *Ocean Circulation*, 2nd edn (Oxford: Butterworth-Heinemann, 2001).

J. T. Houghton, *Global Warming: The Complete Briefing* (Cambridge: Cambridge University Press, 2004).

J.-F. Minster, *La Machine Ocean* (Paris, France: Flammarion, 1997).

J. Merle, *Ocean et Climat* (Paris, France: IRD Editions, 2006).

B. Voituriez, *Le Gulf Stream* (Paris, France: Editions UNESCO, 2006).

7

Ice and climate

RAYMOND C. SMITH AND FRÉDÉRIQUE RÉMY

Generations of men establish a growing mastery over the earth, but they are destined to become fossils in its soil.

– W. and A. Durant

Vous connaissez l'Angleterre; y est-on aussi fou qu'en France? C'est une autre espèce de folie, dit Martin. Vous savez que ces deux nations sont en guerre pour quelques arpents de neige vers le Canada...

– Voltaire, *Candide ou l'optimisme*

The cryosphere

The cryosphere is 'that part of the earth's crust and atmosphere subject to temperatures below 0 °C for at least part of each year.' Here, the cryosphere designates the snow and ice components of the Earth, i.e. sea ice, glaciers, large ice sheets, or seasonal snow cover. The cryosphere occupies a wide range of spatial and temporal scales: spatially, from km to continental in size; temporally, from season to longer than millennia. Except for glaciers at high altitude, the cryosphere is primarily confined to the polar regions.

Continental ice is the largest storehouse of the Earth's freshwater. The Antarctica ice sheet, with a surface of 15 million square km covered by an average of

> **Raymond C. Smith**, ICESS and Department of Geography, University of California, Santa Barbara, is a specialist in ocean optics and until recently was the lead principal investigator for the Palmer Long-term Ecological Research program, including sea ice studies, in Antarctica. He was the founding Director of ICESS and recipient of the Jerlov Award from The Oceanography Society.
>
> **Frédérique Rémy**, from CNRS, is head of the Glaciology Team at LEGOS (Laboratoire d'études en Géophysique et Océanographie Spatiale) in Toulouse. She is a specialist in remote sensing of polar caps and glaciers.

2000 m of ice, contains 90% of the Earth's ice, comprising a storage of 30 million cubic km of ice. The Greenland ice sheet is 10 times smaller, with storage of 3 million cubic km of ice. All the continental glaciers contribute less than 1% of the Earth's ice but are very numerous and well distributed over the Earth, so that they represent a good global indicator of climate variability.

In the Arctic, sea ice typically covers about 14–16 million square km at its maximum extent in late Northern Hemisphere winter and 7–9 million km^2 at its minimum seasonal extent in late summer. In the Antarctic, sea ice is at its maximum during austral winter and covers 17–20 million square km, while covering only about 3–4 million square km at its minimum during the Southern Hemisphere summer.

Polar regions are those components of the global climate system, whereby heat gained at low latitudes is balanced by heat loss at high latitudes via ocean and atmospheric circulation. The sea ice plays a substantial role in the global climate via strong feedback systems, both positive and negative, while ice sheets and glaciers are particularly significant for the storage of freshwater. We consider these roles in more detail below.

A close relationship between ice and climate

Ice and snow are closely linked with climate, both respond to climate change over a wide range of temporal scales, and both, in turn, act on climate.

First, sea ice is both an indicator and, via strong feedback mechanisms, an agent of climate change. Sea ice, in the context of the global climate system, mediates between ocean and atmosphere by regulating the heat, moisture, salinity, and surface albedo and, consequently, influences both global ocean and atmospheric circulation. Long-term variations in sea ice coverage are indicative of large-scale and long-term climate change, while short-term variations in sea ice reflect variability in various climate oscillations and modes of atmospheric variability. On a regional scale, seasonal sea ice advance and retreat strongly influence the physical and biological environment. In turn, these seasonal changes significantly feed back on the global climate system as well as influence surrounding ecosystems, where sea ice provides habitats of major ecological importance. Because the potential impacts of global warming on sea ice are so substantial in terms of climate, polar feedback systems, ecosystems, and social consequences, sea ice is an important variable in any assessment of climate change.

Second, continental glaciers and ice sheets are vast ice reservoirs whose volume is controlled by the balance between snow accumulation and loss by melting or ice flow. In the case of mountain glaciers, snow mostly falls at high altitude in the so-called accumulation zone, transforms into ice as it sinks, and flows in the downstream direction to lower altitudes, where the temperature is such that the loss by melting is larger than the snow gain. Ice flow that redistributes ice mass from top to bottom is thus the cause of the presence of glaciers in the 'green valleys' that so astonished the naturalists of the eighteenth century. In some cases; e.g. the large ice sheets of Greenland and Antarctica, the flow reaches the sea and loss occurs by calving. The ice loss for continental glaciers is mostly due to melting, while, for the Greenland ice sheet, one-half of the ice loss is due to iceberg calving and the other to melting. By contrast, the climate in Antarctica is such that the loss is almost all due to iceberg calving.

Snow precipitation, snow melting, and glacial ice dynamics are strongly dependent on climate. First, ice dynamics adapt to the geometry of the glacier, surface slope, and ice thickness, such that fluctuations induced by changes in the climate forcing play a role in how the glacier evolves. Second, the speed and volume of ice flow increase with temperature. However, temperature change penetrates very slowly into the base of the glacier where deformation occurs, so that the response is delayed. For instance, the heat wave that occurred 15 000 years ago has not yet reached the deepest layers of the Antarctic ice sheet.

Conversely, the response of precipitation and snow melting to climatic variations is relatively fast, on the order of years to decades. As a consequence, ice dynamics response varies from a few years for a continental glacier to tens of thousands of years for the Antarctic ice sheet. Thus, the health of the glacial reservoirs depends not only on the current climate but also on its history, from decades to millennia.

Outlet glaciers classically have been presumed to respond slowly to climate change, with response times of centuries to millennia. Several investigations, however, are suggesting that warming may have an almost immediate influence on glacier speeds and the subsequent discharge of ice into the ocean. As a consequence, global warming may increase the contributions of Antarctica and Greenland to sea level rise faster than previously anticipated.

Recent observations

In the last four decades, remote sensing by satellite has offered a new vision of the evolution of the cryosphere. Indeed, accessibility, geographic size, and harsh climatic conditions of sea ice, glaciers, and ice sheet strongly limit *in situ* observations. Observations from microwave radiometers have permitted sea ice extent (a measure of the area that contains at least 15% sea ice) and areal coverage (actual area covered by sea ice) to be monitored since the 1970s. Studies using these data have shown the dynamic nature of sea ice cover and allowed characterization of its varied spatial and temporal response to climate forcing.

Remote sensing of ice sheets really began in 1991 with the launch by the European Space Agency of ERS-1, a polar-orbiting satellite with instruments adapted to polar observations. In particular, a radar altimeter initially devoted to oceanic survey provided time series of continental ice surface height, and a synthetic aperture radar provided information on ice deformation and flow. Recently, new measurements of time-variable gravity from the Gravity Recovery and Climate Experiment (GRACE) satellites have also been used to estimate the loss of Antarctic ice mass.

Remote sensing of continental glaciers is still limited, and the best space-based tool remains the imagery from high-resolution optical sensors, such as those offered by the SPOT series. The effective retreat of continental glaciers and that expected of ice sheets are now becoming well documented.

Sea ice

Satellite observations of sea ice from passive microwave radiometers have also been combined with earlier observations from ice charts and other sources to yield a time series of Arctic sea ice extent from the early 1900s onward. Ice thickness is an equally important parameter for assessing sea ice in the context of an ice mass budget but is more poorly determined. Ice thickness, its spatial extent, and the fraction of open water within the ice pack can vary rapidly over large spatial scales in response to wind, currents, and seasonal variability. Currently, sea ice thickness observations are relatively restricted in space and time. Holloway and Sou [46] provide a review and analysis of Arctic Ocean ice thickness observations and conclude that the volume loss from 1987–97 lies within the range of 16%–25%. Based on the analysis of submarine-based sonar profiling data and consistency among sea ice models, Rothrock *et al.* [47] suggest that the ice cover over the central Arctic Ocean has thinned during the period of 1987–97. There is less information with respect to the thickness of seasonal sea ice around the periphery of the Arctic Ocean. We know of no systematic large-scale observations of ice thickness for the Southern Ocean.

During the recent satellite era, passive microwave data reveal that Arctic ice extent decreased by about 3% per decade, while Antarctic ice extent increased by 0.8% per decade. However, individual regions within the polar oceans often display individual trends. For example, the western Antarctic Peninsula region has undergone a warming in winter of almost 6.0 °C (10.7 °F) since 1950, a loss of seven ice shelves, the retreat of 87% of the marine glaciers, and a decrease in winter sea ice duration of 30–40 days during the past few decades, all of which have significantly impacted the marine ecology of this area.

Within the last few years, satellite data have indicated an even more dramatic reduction in Arctic Ocean ice cover. In particular, September sea ice during the

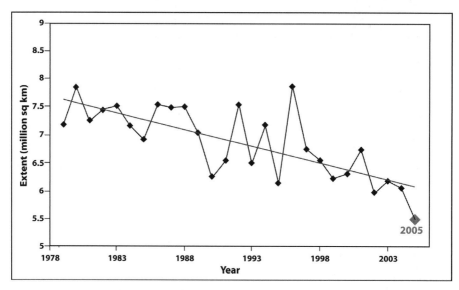

Figure 7.1 Decline in Arctic sea ice extent from 1978–2005. The September trend from 1979–2005, now showing a decline of more than 8% per decade, is shown with a straight line (after F. Fetterer and K. Knowles, 2002, updated 2005 [48]. National Snow and Ice Data Center, Boulder, Colorado).

past few years has set successive records for minimum sea ice extent (Figure 7.1). These recent data show the September Arctic ice extent trend for 1979–2005 to have declined by more than 7.7% per decade [48]. The Arctic Ocean ice cover is now at its lowest level in more than a century and suggests that 'Arctic sea ice extent is also beginning to show the signature of anthropogenic climate change.' In another report, Overpeck *et al.* [49] suggest that the 'Arctic system is moving toward a new state that falls outside the envelope [...] of recent Earth history' and state that '[t]he change appears to be driven largely by feedback-enhanced global warming.'

Even though sea ice occurs primarily in polar regions, it influences global climate through various feedback mechanisms, several of them being strongly positive. The albedo of sea ice is high, especially if snow covered, so that a high fraction of incident solar radiation is reflected back into space. Consequently, snow/sea ice–covered areas absorb relatively less solar energy than do lower-albedo land or water, and temperatures in the polar regions remain relatively cool. Global warming reduces the snow/ice cover, reducing the albedo of the affected area and permitting more solar radiation to be absorbed in the now ice/snow free areas, raising further the temperature. This cycle of warming and melting creates a strong positive feedback, further destabilizing the sea ice pack. This positive feedback mechanism (surface albedo feedback), whereby a small initial temperature

Glaciers and ice sheets

Who hasn't heard a mountaineer complain about the retreat of the glaciers? Indeed, a general retreat of glaciers is observed in the Alps, Himalaya, Andes, North America, Arctic, or Patagonia. However, it should be noted that only a few tens of glaciers of 160 000 are well documented from field observations. Among these, the Alpine glaciers have been studied since the beginning of the twentieth century. They exhibited a first retreat in the 1940s after successive winters, with limited snowfalls combined with warm summer seasons, and a second episode since the 1980s, clearly due to an increase of the snow/ice melting. Remote sensing now allows for the monitoring of a large number of glaciers and the estimation of volume and surface velocity changes. In many instances, the increase of melting cannot solely explain the current pace of glacier thinning. A dynamical response may be also invoked. Since 2000, glacier melting contribution to rising sea level is estimated to range between 0.6 mm and 0.8 mm per year, twice as much as their mean contribution between 1950 and 2000. The most sensitive glaciers are the calving ones, especially the glaciers from Patagonia, whose ice loss contributes more than 10% of the global total. It is widely accepted that this general glacier retreat will increase during the next decades.

Observations of the Greenland ice sheet topography, with the help of radar altimetry, clearly show a thickening of the central part and a thinning of marginal parts, especially on the southeastern coast, with the total volume remaining fairly constant. For a long time, this Greenland 'deformation' has been considered an academic example of the predicted parallel increases of snow precipitations in the central region and of melting in coastal areas. However, recent radar observations show that the thinning of the marginal zone is also due to an increase of outlet glacier flows and thus to the beginning of a dynamical response. In the long term, this suggests an acceleration of imbalance for this ice sheet, i.e. the ice sheet will decrease. Currently, the balance of the Greenland ice sheet is close to equilibrium, but it may contribute to sea level rise in the near future if the loss of ice mass resulting from enhanced glacier melt and flow increases.

The first worrisome observations about the Antarctic ice sheet were the acceleration of emissary glaciers in a small sector of the West Antarctic (Figure 7.2) and the collapse of large floating ice in the peninsula sector. Indeed, in the central part of the Antarctic, snow precipitation, ice flow, and climate changes are so weak that the identification of any climatic trend with any confidence is still difficult. However, the Western part is found to lose mass at a rate of 45 Gt per year

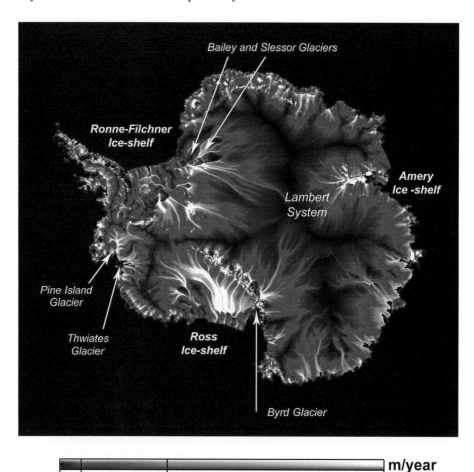

Figure 7.2 Ice flow in Antarctica: one can observe large ice flows still visible several hundreds of km upstream. Eighty percent of the continental ice are drained by only a few percent of the coast. (For image in color, please see Plate 7.)

due to the long-term response to the last climatic warming. This loss is far from being compensated by the ice mass gain of 15 Gt per year of the Eastern part that results from a slight enhancement of snow precipitation. The total contribution of the Antarctic to sea level rise, as currently measured by altimetry, seems to be negligible. The contribution recently estimated from the GRACE mission suggests a more pronounced decrease of the Western part, but these results are still not fully confirmed.

Recent paleoclimatic studies suggest that 'polar warming by the year 2100 may reach levels similar to those of 130 000 to 127 000 years ago that were associated with sea levels several m above modern levels; both the Greenland Ice Sheet and portions of the Antarctic Ice Sheet may be vulnerable. The record of past ice-sheet melting indicates that the rate of future melting and related sea level rise could be faster than widely thought.'

Possible mechanisms influencing observed changes in the cryosphere

The surface (snow/ice) albedo feedback, discussed above, plays an important role in the amplification of polar temperatures. Polar amplification refers to the amplification of surface temperature changes in polar regions compared with the rest of the globe in response to a change in climate forcing, either natural or anthropogenic. For example, there is agreement among most models that the Arctic warms more than subpolar regions when subjected to increasing levels of greenhouse gases (GHG) in the atmosphere. This is in contrast [3] to the Antarctic, where warming is retarded (except in the Antarctic Peninsula region) by the large heat uptake of the deep circumpolar Southern Ocean and anomalous atmospheric circulation. Recent work shows that Antarctic surface temperature cooling is consistent with circulation changes associated with a shift in the Southern Annular Mode[1] (SAM) [50]. The authors suggest that both Antarctic ozone depletion and increasing greenhouse gases have contributed to the observed trends. They further suggest that Antarctic cooling may reverse when the effects of forcing due to ozone depletion begin to recover in the middle of this century, leading to a GHG warming pattern.

Another positive feedback associated with snow cover is the deposits of black soot due to pollution over snow surface, which decreases the ice albedo – all the more as snow melting concentrates the soot density. Consequently, decreased snow albedo leads to enhanced warming.

Holland and Bitz [51], evaluating results of 15 fully coupled atmosphere–ocean–sea ice–land models, have estimated the polar amplification for snow coverage on land and found that it is weakly related to snow extent and sea ice extent. They conclude that snow coverage on land 'has little influence on the simulated Northern Hemisphere polar amplification' and that its influence is primarily a consequence of the relationship 'between ice extent and snow cover.'

Thermohaline feedback is another potential feedback of concern within the context of climate change. As increased warming leads to the export of increased freshwater from the Arctic Ocean, stratification of the North Atlantic is hypothesized to follow, slowing down the thermohaline circulation [52]. Changes in poleward ocean heat transport influence the sea ice cover and the heat available to the atmosphere and thus contribute to polar amplification (see Chapter 6).

---

[1] The Southern Annular Mode (SAM), also referred to as the Antarctic Oscillation, is the dominant pattern of nonseasonal tropospheric circulation variations south of 20 °S, and it is characterized by pressure anomalies of one sign centered in the Antarctic and anomalies of the opposite sign centered about 40–50 °S.

Cloud cover changes at high latitudes, accompanying global warming, have been shown to have both positive and negative feedback effects. It has been shown that models with large winter cloud cover experience positive feedback and higher polar amplification, presumably because of changes in the downward surface longwave radiation.

Corell (see Chapter 9) discusses the role of methane as a potential greenhouse gas. The release of methane, with increasing global temperatures, creates a positive feedback process, leading to further increases in global temperatures.

The studies summarized above, among many others, emphasize the complexity of polar amplification and the various processes that contribute to this amplification. Analyses of various Arctic forcings and feedbacks show that the amplification arises from 'a balance of significant differences in all forcings and feedbacks between the Arctic and the globe.' Because of all these positive feedbacks, we can predict an acceleration of ice loss and a subsequent sea level rise.

Priority issues for research

The evidence for global warming in polar regions is considerable: retreat of polar glaciers, acceleration of marginal ice sheet flow, large floating ice collapse, decrease of both sea ice extent and thickness, and ecosystem response to these climate changes.

Furthermore, models, paleoclimatic records, and instrumental data were consistent in showing that Arctic surface air temperatures in the twentieth century were exceptionally high compared with, at least, the previous 300 years. Arctic sea ice extent and polar glacier volume have shown a corresponding decline during the last century and certainly during the satellite era of the past three decades. Also, it is well established that climate models can reasonably simulate past records and that they exhibit polar amplification associated with future climate change scenarios. However, there are several critical areas where increased data, modeling efforts, and understanding are still required in order to simulate more accurately key patterns of variability (e.g. Arctic Oscillation, Southern Annular Mode, etc.) and to make accurate prediction of future climate change.

These include:

1. better data – especially in the Southern Hemisphere and concerning ice sheet mass balance;
2. a more complete understanding of the mechanisms of polar amplification and the various feedback systems influencing this amplification at high latitudes;
3. a better understanding of tidewater outlet glaciers of the Greenland and Antarctic ice sheets and the role of subsurface oceanic waters in melting the submarine bases of these glaciers;

4. a more complete modeling of ice sheet dynamics and of internal feedback, such as surge or floating ice collapse, and incorporation of important physical processes implied in paleoclimatic analyses that are currently not included in ice sheet models; and
5. a better understanding of how polar ecosystems are responding to the recent rapid warming.

There is also a need, within the context of indigenous communities, to document existing knowledge retained by elders before this knowledge is lost as the elders pass away and the environment changes.

Societal implications

Glaciers play a fundamental role in water resource regulation, smoothing torrents and river flow at the seasonal scale. More than one billion people depend on glacier water for food, agriculture irrigation, or hydropower resources. Until now, water flow from glaciers has been increasing, but it is predicted to decrease or even disappear in the long term. Vast regions of Himalaya, Andes, or China could face serious water resource problems in the near future.

Most importantly, the retreat of glaciers and ice sheets significantly contributes to sea level rise (around 1 mm per year), whose societal implications are quite serious (see Chapter 6).

The Arctic Climate Impact Assessment (ACIA 2004) provides a detailed review [28] and analysis of climate change impacts on the Arctic, including societal implications. The ACIA report (see Chapter 9) reviews the collective knowledge of indigenous populations, summarizes their observations, and examines specific perspectives of several communities or peoples. An important observation is that the weather, as distinct from longer-term climate change, has become more variable and hence less predictable with traditional methods. Many people report that the recent climate and environmental changes are outside the variability expected based on long-term experience. Issues reviewed and discussed include increased solar ultraviolet radiation due to decreased high-latitude ozone concentrations and the influence on indigenous peoples; changes in near-shore sea ice extent and thickness and the negative impact of these changes on traditional hunting practices; various impacts on the ecosystems on which these peoples exist; and the impacts on humans and their way of life.

Conclusions

Glaciers are retreating, the large Greenland and West Antarctic ice sheets are melting, ice shelves are collapsing, Arctic sea ice is shrinking, and snowmelt is occurring earlier. In 2007, sea ice reached a record low and the September decline trend is now over 10% per decade [54]. All of these, through positive feedbacks, are increasing the surface warming rates. The recent collapse of the Larsen B ice shelf in Antarctica that is 'unprecedented during the Holocene' is but one confirmation that polar regions are warming faster than lower latitudes. Furthermore, all recent evidence suggests that the polar cryosphere is changing much faster than previously anticipated. In particular, glacier discharge from tidewater outlet glaciers of Greenland and Antarctic ice sheets has increased significantly in the past decade. It is hypothesized that this acceleration is due to warm water from intermediate depths melting the floating ends of the glaciers from below. This is a process that will accelerate with global warming.

Critical consequences of adding freshwater from ice sheet melting to the oceans include a slowing of ocean thermohaline circulation, with subsequent impact on global weather patterns and a rise in sea level. A recent estimate suggests that by the end of the twenty-first century, sea level is projected to rise by 0.5 ± 0.4 m in response to additional global warming. The potential contribution from the Greenland and Antarctic ice sheets dominates the uncertainty in this estimate.

In snow-dominated regions, climate change is altering the hydrological cycles. Regions where the supply of freshwater is currently dominated by melting snow or ice are especially vulnerable, with potential long-term implications for food availability and the viability of local ecosystems. Barnett and co-workers [28] note that 'more than one-sixth of the Earth's population relies on glaciers and seasonal snow packs for their water supply,' and they suggest that the consequences of hydrological changes to these populations are likely to be severe, including loss of potable water, agricultural disruptions and/or losses, possible displaced populations, and extensive changes to ecosystems and the loss of services that humans derive from them.

Polar ecosystems are being, and will continue to be, significantly impacted by global warming. Temperature is a critical climatic factor influencing polar ecosystem structure and function because it directly influences the phase of water. Relatively small variations in temperature, cloudiness, and solar radiation determine the number of days above freezing and thus the availability of liquid water. In turn, availability of liquid water is an important driver for the function and biodiversity of many polar ecosystems. For marine ecosystems, relatively small temperature changes give rise to glacier melt, ice shelf collapse, and sea ice reduction, with subsequent impacts at all trophic levels within such an ice-dominated system. It is important to recognize that the ice-to-water phase transition is, therefore, a critical temperature threshold that may induce relatively large nonlinear ecological responses to a relatively small temperature change. Consequently, these polar systems are extremely sensitive to climate variability.

8

The changing global carbon cycle: from the holocene to the anthropocene

BERRIEN MOORE III AND PHILIPPE CIAIS

When one tugs at a single thing in nature, he finds it attached to the rest of the world.

– John Muir

Les forêts précèdent les peuples, les déserts les suivent.

– François-René de Chateaubriand

Introduction: dust to dust

Planetary atmospheres with an incident solar energy flux are subject to an energy pumping that takes the form of a matter flux from lower chemical potential

Berrien Moore III joined the University of New Hampshire (UNH) faculty in 1969, soon after receiving his PhD in mathematics. He was named University Distinguished Professor in 1997. He has led the Institute for the Study of Earth, Oceans and Space at UNH as Director since 1987. He has been a visiting scientist in several institutes in the United States and abroad, including the Laboratoire de Physique et Chimie Marines at the Université de Paris. Professor Moore chaired the Scientific Committee of the International Geosphere-Biosphere Programme (IGBP). He has authored over 150 papers on the global biogeochemical cycles and global change, as well as numerous policy documents in the area of the global environment. He has chaired NASA's senior science advisory panel, was a member of the NASA Advisory Council, and has contributed actively to committees at the US National Academy of Science. Currently, he is a member of the Space Studies Board. In May 1992, he was presented with the NASA Distinguished Public Service Medal for outstanding service to the agency.

Philippe Ciais graduated with a degree in physics from École Normale Supérieure de Saint-Cloud in 1989 and wrote his doctoral thesis in isotope glaciology in 1991. After completing a postdoctoral fellowship in 1992–1993 at NOAA in Boulder, Colorado, he joined the Laboratoire des Sciences du Climat et de l'Environnement in Gif-sur-Yvette in 1994, of which he currently is Associate Director. Dr. Ciais has contributed to 96 peer-reviewed articles and several book chapters, including this one. He coordinates several French and European research projects and acted as a lead author of the IPCC.

forms to those at a higher potential. When life is present, this geochemical cycle can take another pathway: the basic chemical constituents of organic matter, carbon, nitrogen, oxygen, phosphorus, and sulfur follow a closed loop or cycle through increasing molecular energy states as the elements are incorporated into living tissue and then decreasing energy levels as the tissues decompose, giving rise to *biogeochemical cycles*. The chemical disequilibrium observed on Earth is a signature of a living planet. It is also a reflection of the ancient phrase 'dust to dust.'[1]

The significance of the role of living systems in all of the Earth's geochemical cycles is a relatively recent discovery. The recognition of biotic controls of biogeochemical cycles and thereby of the climate system (in part because the greenhouse gases are state variables in the biogeochemical cycles) is now central to our understanding of the 'planetary metabolism,' including composition of and controls on the atmosphere, oceans, and sediments on the surface of our planet. The dynamic patterns of these biogeochemical cycles are the consequences of a myriad of processes that operate across a wide spectrum of time- and space scales. In the absence of significant disturbances, these processes define a natural cycle for each element, with approximate balances in sources and sinks that result, at least on timescales of less than a millennium, in a quasi-steady state.

Since ancient times, humans have modified natural systems and thereby affected their local environments. However, since the beginning of the Industrial Revolution, human activity has begun to change the environment on larger and larger scales. The evidence is now overwhelming that human activity has significantly altered biogeochemical cycling at the regional, continental, and planetary scales [6]. These cycles are now far from a quasi-steady state. In effect, humans have modified the planet's metabolic system, and this reality led Paul Crutzen to term the modern era as the Anthropocene Era [55, 56].

For instance, the value of the concentration of atmospheric carbon dioxide (CO_2) and methane (CH_4) have moved into a range without precedent for humans. We know that the carbon pool in the atmosphere in the form of CO_2 has increased from about 590 to almost 780 Gt C (1 Gt C $= 1 \times 10^9$ tons C $= 1 \times 10^{15}$ g C $= 1$ Pg C) between 1765 and 2000 as a result of fossil fuel burning and forest clearing. The annual rate of increase is almost 0.5%. We have a direct record (Figure 8.1) of this increase since 1958. From the ice core records [57, 58], we know that the concentration of CO_2 was relatively constant over the last 1000 years to the onset of the anthropogenic increases (Figure 8.2) in the middle of the eighteenth century.

---

[1] Often used in Christian burial services and based, in part, on Genesis 3:19 (New International Version): 'By the sweat of your brow you will eat your food until you return to the ground, since from it you were taken; for dust you are and to dust you will return.'

The changing global carbon cycle: from the holocene to the anthropocene 123

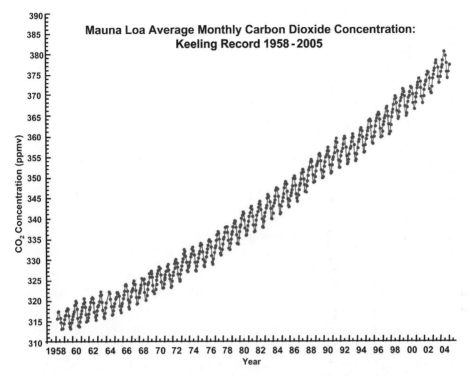

Figure 8.1 Mauna Loa monthly atmospheric carbon dioxide concentration: Keeling record, 1958–2005.

In the same time frame, atmospheric CH_4 has nearly tripled in concentration from 700 ppb in the preindustrial times up to 1900 ppb today [59]. Methane presumably started to rise slightly before CO_2 did, in response to early livestock domestication, rice agriculture, and human-caused fires [60]. The rate of growth of atmospheric CH_4 has decreased recently; we know neither the causes of this pause nor its likely duration.

On longer timescales, as shown by Antarctic ice core records [61, 62, 63], there is an intriguing and periodic signal of change in the carbon cycle (Figure 8.3). The periodicity of interglacial and glacial climate periods are in step with the carbon cycle as significant pools of carbon are slowly transferred from the land through the atmosphere to the ocean as the planet enters glaciation, and then there is the rapid recovery of carbon from the ocean back through the atmosphere and onto the landscape as the planet exits glaciation. The repeated pattern of a 100-ppm decline in atmospheric CO_2 from an interglacial value of 280–300 ppm to a 180-ppm floor and then the rapid recovery to 280 ppm as the planet exits glaciation suggests a tightly governed control system, with firm stops at roughly 280 and 180. There is a similar methane cycle between approximately 350 ppb

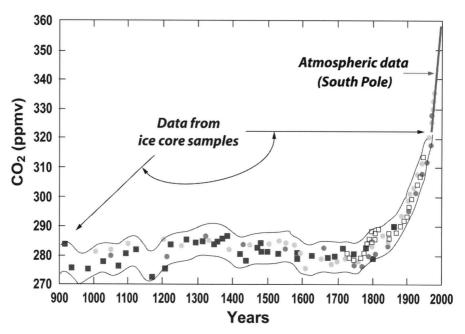

Figure 8.2 Carbon dioxide concentration from ice cores (since 900 AD) and atmospheric data (last 50 years).

and 700 ppb that is also in step with temperature. What begs explanation is not just the linked periodicity of carbon and glaciation but also the apparent hard stops in the carbon system. What were the controls, and why are there the 'hard stops?'

Today's atmosphere, imprinted with the fossil fuel CO_2 signal, stands at nearly 100 ppm above the previous 'hard stop' of 280 (see 'star' on Figure 8.3). Methane is nearly three times its preindustrial level. In essence, carbon has been moved from an immobile pool: the fossil fuel reserves part of the slow, geological carbon cycle to an extremely mobile pool, the atmosphere. In the fast carbon cycle, the ocean and terrestrial vegetation and soils have yet to equilibrate with this rapidly changing load of carbon in the atmosphere. From the perspective of carbon in the atmosphere, we are now in a zone not visited in the last 25 million years. Focusing on the future level of CO_2 compared with its past, the Intergovernmental Panel on Climate Change (IPCC) 92a projections of the next century takes on an expanded meaning. Regardless of possible changes in climate, one cannot help but wonder about the characteristics of the carbon cycle in the future.

The primary human activities contributing to the current change in the global carbon cycle are fossil fuel combustion (Figure 8.4) and modifications of global vegetation through land use (Figure 8.5). Globally, fossil fuel is burned at a rate

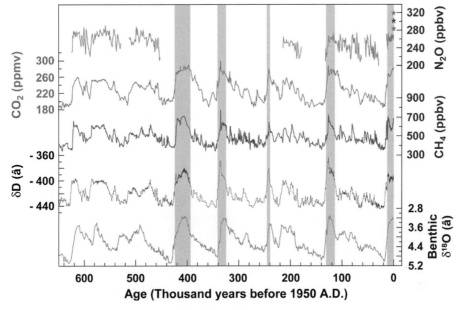

Figure 8.3 Reconstructed variations of carbon dioxide, methane, and nitrous oxide concentrations and of other parameters from Antarctic ice core over the past 650 000 years.

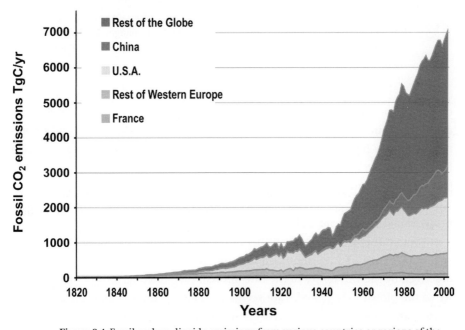

Figure 8.4 Fossil carbon dioxide emissions from various countries or regions of the world.

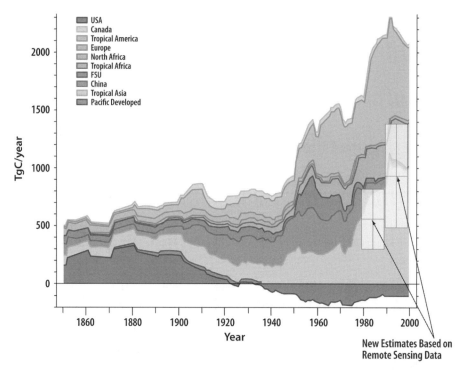

Figure 8.5 Annual amount of carbon released by the continental biosphere due to land use changes in various regions of the world. In a few regions, including the United States, the carbon emission budget from land use changes is negative. Global estimates in the 1980s and 1990s and associated uncertainties based on remote sensing data are indicated by the crossed boxes (after [93]).

of slightly more than a ton of carbon (as CO_2) per person per year – although the country-to-country differences (implicit in Figure 8.4) are significant and important.[2] Yet only half of the CO_2 from fossil fuel combustion and land-use change has remained in the atmosphere. Atmospheric measurements of carbon isotopes [64] and oxygen-nitrogen ratios [65, 66] have shown that the land and oceans have sequestered fossil carbon in approximately equal proportions. However, the proportional balance varies in time and space.

Although it is clear that fossil fuel burning is the primary cause of the atmospheric increase in CO_2, the variability in the rate of that increase cannot be described by the variability in the rate of fossil fuel burning. Rather, the variability of the global CO_2 growth rate seems to reflect primarily changes in carbon sources and sinks in terrestrial ecosystems [67, 68, 69, 70] that are connected with

---

[2] See http://cdiac.esd.ornl.gov/trends/emis/em_cont.html.

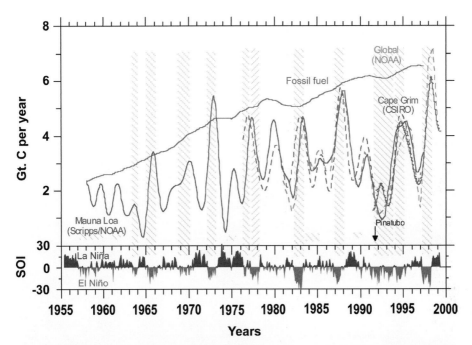

Figure 8.6 CO_2 annual growth rate (1955–2000). The fossil fuel emission growth rate increases smoothly, while the observed CO_2 concentration growth rate shows a high variability. This variability is connected to carbon sources and sinks in terrestrial ecosystems and to large-scale phenomena such as El Niño.

large-scale climate modes of variability (Figure 8.6). The underlying processes are not fully understood. Knowledge of today's regional CO_2 sources and sinks, including their mechanisms and their sensitivity to climate perturbations, is central to theoretical predictions of future levels. However, the geographic distribution of carbon sources and sinks has remained elusive [64, 71, 72, 73, 74, 75]. As nations seek to develop strategies to reduce their carbon, the capacity to quantify and understand the present-day *regional* carbon fluxes [76, 77, 78] is a prerequisite to prediction and, thereby, informed policy decisions. This does not imply, however, that all policy actions need to remain on hold.

The increase in the atmospheric CO_2 concentrations (as well as other greenhouse gases) due to human activity has produced concern regarding the heat balance of the global atmosphere. Specifically, the increasing concentrations of these gases will lead to an intensification of the Earth's natural greenhouse effect, and this shift in the heat balance will force the global climate system in ways that are not well understood, given the complex interactions and feedbacks involved. There is general agreement that global patterns of temperature and precipitation

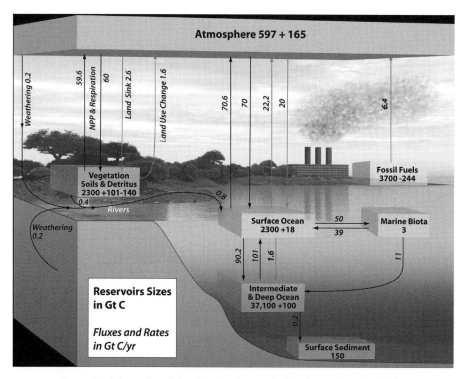

Figure 8.7 Schematics of the global carbon cycle. (For image in color, please see Plate 8.)

will change, although the magnitude, distribution, and timing of these changes are far from certain (see Chapters 1, 2, and 4).

Understanding the carbon cycle (Figure 8.7) and its underlying processes is a key to comprehending the changing terrestrial biosphere, understanding the role of the oceans in planetary biogeochemistry, and developing a reasonable range of future concentrations of carbon dioxide and other greenhouse gases. Conversely, our predictions about the physical climate system and climate change are confounded by the fact that the carbon cycle is still not adequately understood or quantified globally.

The ocean system: a big carbon processing machine

The oceans are without question a sink for anthropogenic CO_2 [e.g. 78, 79, 80], but the strength of this sink is still unclear. The primary controls are the circulation of the ocean, primarily the global thermohaline 'conveyor belt' (see Chapter 6 for a discussion of this idea), and two important biogeochemical processes (e.g. [81]): the solubility pump and the biological pump.

The changing global carbon cycle: from the holocene to the anthropocene 129

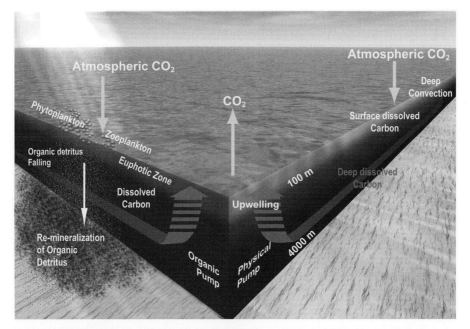

Figure 8.8 Ocean portion of the carbon cycle.

The thermohaline circulation is obviously important because it transports heat from the tropics to the poles (hence its name) and in so doing plays an important role in the oceanic uptake of carbon dioxide, both through the downward flux of water in certain high-latitude regions and the change in the solubility of CO_2 in seawater as a function of temperature. The solubility pump is a lift that carries carbon down from surface to the abyss. The governing dynamics are essentially two-fold: CO_2 is more soluble in cold water than warm, and in certain cold-water regions there is convective overturning (part of the 'conveyor belt') as this water sinks to depths.

The biological pump is a 'pipe' that also brings carbon down to the abyss, but it involves the opposing effects of organic matter and calcium carbonate formation by algae. The formation of organic matter by algae in presence of sunlight withdraws CO_2 from the top of the ocean and hence from the atmosphere. The particles of dead algae sink to the depth, where they decompose. However, unlike the organic matter production, the formation of carbonate skeletons by algae raises the amount of CO_2 dissolved at the surface, which decreases the ocean carbon uptake of atmospheric CO_2.

The interplay (Figure 8.8) between the circulation of the oceans and the two 'pumps' determines the uptake of CO_2 by the ocean. Computer models, which link the physical, chemical, and biological dynamics of the ocean carbon cycle, provide a means for estimating the relative strengths of the processes that control ocean uptake of CO_2 as well as their potential future patterns. One test of these computer models of the ocean carbon machine suggests that if there were no algae, the atmospheric CO_2 concentration before the industrial revolution would have been 450 ppm instead of 280 ppm. State-of-the-art models of the natural ocean carbon cycle currently include the effects of marine biology. However, most assessments of the oceanic uptake of anthropogenic CO_2 have assumed that the biological pump would not be affected by climate change and therefore have only modeled the physical processes [e.g. 79, 82]. More recently, a number of new studies [83, 84] have added the plankton to the ocean chemistry and physics, showing possible ways in which the marine biology might be affected by climate change over a 200-year timescale and conversely. The main conclusion was that, because of the complexity of biological systems, it is not yet possible to say whether some of the likely feedbacks would be positive or negative.

Simulating the carbonate skeletons of algae with a computer model of the ocean carbon cycle requires another level of complexity beyond simulating the organic matter production: the distribution of particular plankton species (mainly coccolithophorids) must be simulated. This is very difficult, so it is 'fortunate' that it appears that the biological carbonate formation contributes relatively little to the vertical carbon gradient compared with the organic matter production. The importance of changes in ocean carbonate continues to require careful evaluation. In particular, on timescales beyond a century, the pool of seafloor $CaCO_3$ sediments becomes an important player in controlling the amount of oceanic dissolved carbon, and its role in the glacial to interglacial carbon cycle changes must be considered [85, 86].

The land system: a fast carbon-processing machine

Atmospheric CO_2 goes into the land carbon cycle by photosynthesis and comes out by the respiration of plants, the decomposition of dead organic matter by microbes in the soil, or combustion.

Looking at photosynthesis–decomposition as a linked process, one sees that some of the carbon fixed by photosynthesis and incorporated into plant tissue is perhaps delayed (from months to hundreds of years, Figure 8.9) from returning to the atmosphere until it is oxidized by microbes or emitted to the air during fire. This slower carbon loop through the terrestrial component of the carbon cycle affects the rate of growth of atmospheric CO_2 concentration and, in its shorter-term expression, imposes a seasonal cycle on that trend (see again Figure 8.1). In the longer term, the carbon cycle is bound up with ecosystem

The changing global carbon cycle: from the holocene to the anthropocene 131

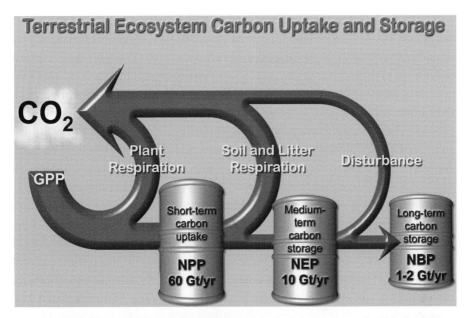

Figure 8.9 Land portion of the carbon cycle.

structure and type. The structure of the vegetation is determined on long timescales by the integrated response to changes in climate and on intermediate timescales by carbon-nutrient machinery. In turn, vegetation feeds back to the climate system, because the plants set up the reflection of solar photons at the surface (albedo) and the transfer of energy and water vapor.

Modeling the complex set of interactions between terrestrial and atmospheric systems requires ecological models (dealing with plants and ecosystems migrations and structure), biogeochemical models (dealing with fluxes of carbon, nutrients, and energy inside ecosystems) and physiological models that describe the turbulent fluxes of CO_2 exchanged with the atmosphere, especially photosynthesis. At each step toward longer timescales, the climate system integrates the finer-scale processes and applies feedbacks onto the terrestrial biome. At the finest timescales, the influence of temperature, radiation, humidity, and winds has a dramatic effect on the ability of plants to transpire and carry out photosynthesis. On longer timescales, integrated weather patterns regulate biological processes, such as the timing of leaf emergence or excision, uptake of nitrogen, and rates of organic soil decay and turnover of soil nitrogen. The effect of climate at the annual or multiannual scale defines the net gain or loss of carbon by the biota [87, 88], its water status for the subsequent growing season, and even its long-term ability to survive in the presence of infrequent climate extremes (droughts, windstorms, etc.) that sporadically ravage ecosystems (e.g. [89]).

As the temporal scale is extended, the development of computer models of the vegetation dynamics, which respond to climate and human land use as well as to other environmental changes, is a central issue. Such models must capture patterns of plant growth, controlled by processes such as reproduction, establishment, and light competition, which have been added to models interactively with the carbon, nitrogen, and water cycles. For instance, the recovery of natural vegetation in abandoned areas depends on the intensity and length of the agricultural activity and the amount of soil organic matter on the site at the time of abandonment. Disturbance regimes such as fire also need to be included in order to successfully treat competitive dynamics and hence future patterns of ecosystem distribution, composition, productivity, and carbon exchange.

This coupling across a range of timescales represents a significant challenge. Immediate challenges that confront models of the terrestrial-atmosphere system include exchanges of carbon and water between the atmosphere and land and the terrestrial sources and sinks of trace gases. Ultimately, adequately addressing such changes will require a much clearer understanding of the nitrogen and phosphorus cycles that intervene in the carbon cycle and can limit plant growth. Finally, there is an added complexity: humans interact directly and actively with the terrestrial carbon cycle through land use, and the patterns of this activity can affect terrestrial carbon systems on timescales from days to centuries.

Land-use changes: clearing the tropical forests to death

Human activities have altered the Earth's vegetation cover in many parts of the world. These changes in land cover can have profound impacts on environmental systems around the globe – including the links between land, water, and air. They also alter the global carbon cycle. Long before the Anthropocene Era, widespread deforestation had taken place in the Old World. Our understanding of these historic land-use changes is anecdotal or of a very local nature; however, it is clear from the ice-core records that the carbon cycle was not significantly perturbed. During the Anthropocene, the rapid human population growth combined with increasing resource consumption greatly accelerated human-induced changes of the Earth's land cover [90, 91]. By 1990, more than one-third of the land surface was used in agriculture: approximately 1471 million ha was used for cropland and an additional 3451 million ha was used in pasture [92].

Since 1700, forest clearing to establish cropland was the main category of land-use change in Eurasia and, later, in the first half of the twentieth century in North America [91]. In recent decades, net forest clearing has nearly stopped in northern regions, and forests are even gaining area in Western Europe and in the United States; however, by the 1960s, deforestation had moved to the tropics. Yet, in a recent report in the Proceedings of the National Academy of Sciences by Kauppi and others,[3] there are now hopeful signs globally. Amid widespread reports of deforestation, some nations have nevertheless experienced transitions from deforestation to reforestation. Among 50 nations with extensive forests reported in the Food and Agriculture Organization's comprehensive Global Forest Resources Assessment 2005, no nation where annual per capita gross domestic product exceeded $4600 had a negative rate of growing stock change.

The net flux of carbon from land-use activities is the sum of carbon emissions (including on-site burning and/or slash and soil organic matter decomposition and the oxidation of harvested products) due to land conversion and logging, and carbon uptake on lands recovering from prior land-use activities. Due to the difficulty in measuring these terms (including state variables like biomass and rates such as deforestation), estimates of the net land-use carbon flux bear a large uncertainty. The land-use induced flux of CO_2 to the atmosphere during the 1990s is estimated to be 1.6 (0.5–2.5) Gt C per year (mean and error range of estimates in [90, 93, 94] as seen in the IPCC Fourth Assessment Report [5]). The global net terrestrial carbon sink based either on the trends of oxygen and carbon isotopes or deduced[4] from the (more robustly estimated) ocean uptake is -1 ± 0.5 Gt C per year during the 1990s. Assuming the land-use emission of CO_2 of 1.6 (0.5–2.5), this leaves a residual land sink of −2.6 (−4.3 to −1) Gt C per year for the 1990s.

On the other hand, new estimates of the area of land recovering from prior land-use activities suggest that 10–44 million km^2 of land are recovering, about half of which is forested [95]. The net regrowth of secondary forests and forest soils over an area of this size could be on the order of 1 Gt C per year and hence lead to substantially lower global estimates of net CO_2 release from land-use activities. Over the coterminous United States, the net regrowth of secondary forests and forest soils from land-use changes per se is the dominant contemporary carbon sink mechanism [96, 97]. Historical model reconstructions illustrate that the presence of this sink is due to net emissions from land-use changes in the past and that this sink mechanism will not scale with CO_2, but instead will decline in the future as ecosystems mature.

Determining the effects of land-use and land-cover change on the carbon cycle depends on an understanding of past practices, current patterns, and projections of future land-use and -cover as affected by human institutions, population size and distribution, economic development, technology, and other factors. The combination of climate and land-use change may have profound effects on the habitability of Earth in more significant ways than either acting alone.

---

[3] See www.pnas.org/cgi/content/full/103/46/17574
[4] Essentially, treating the land-use induced flux as the residual in fossil fuel burning, ocean uptake, and atmospheric concentration.

A particularly interesting and dynamic process is the human use of fire as an agent in land-use change.

Fires: between Nature's necessity and human pressure

Wildfires are a key process in the global carbon cycle. They affect nearly all types of ecosystems and release, in a very brief period, large amounts of CO_2 (and other reactive carbon gases that enter dynamically into the chemistry of the atmosphere), carbon that was accumulated slowly in the ecosystem. 'Slow in, fast out' is the governing dynamic of carbon in ecosystems affected by fire. From a carbon perspective, the release of carbon from fires is subsequently offset by carbon gains during periods of regrowth following fires; trading time for space, we see that over large spatial and temporal scales, these large but dynamically different gross fluxes tend to compensate and lead to smaller net fluxes.

Fire frequency has dramatically increased since humans started setting fire to ecosystems for hunting and later for deforestation and agriculture [98]. At present, the burning of ecosystems is happening mainly in the tropics. Satellite observations indicate that today, 70% of detected fires are in the tropics, with 50% in Africa [99]. The majority of the lands affected by fires are woodlands and scrublands, and thus the impact on the carbon cycle is less significant than that of forest fires. Closer to the Equator, rainfall is more abundant, and fewer seasonal tropical rain forests, which have the highest aboveground biomass content in the world, are found [100]. Because of the prevailing wetness, these forests are only seldom subject to fire. However, they can burn during severe El Niño–induced droughts. In the past decades, the occurrence of fire has also increased in the tropical moist forests of South America [101], Africa [102], and Asia [103] in response to large-scale deforestation and logging as roads penetrate remote regions.

The process of mechanized and widespread deforestation, in which forests (high carbon stocks) are cleared by burning and replaced by 'relatively permanent' agricultural plants (low carbon stocks), is the largest human-induced (not counting fossil-fuel burning) terrestrial source of CO_2.

The burning of peat deposits in tropical forest regions has received more attention lately. These peat layers have accumulated carbon for thousands of years but are now being drained (to enable agriculture), and this leads to fire susceptibility. In these peat layers, the fuel loads are high and far exceed those of the aboveground vegetation in the forests. In the exceptionally large El Niño of 1997–1998, peat fires in Indonesia emitted as much CO_2 as half of fossil emissions [104].

Fire is the main source of disturbance in boreal forests and has return frequencies between 50–200 years [105]. Fire is part of a natural life cycle in which vegetation succession stages lead to old growth conifer forests, which are especially prone to fire. Although the flames kill the trees, the vegetation is well adapted to a fire; black spruce, for example, will only open its cones to disperse seed to the forest floor to regenerate during a fire. Boreal conifer forests in Canada and Siberia contain large amounts of carbon, especially belowground (75 Gt C in forests; 250 Gt C in soils for Siberia alone). Certain years are 'big fire years'; generally, these are dry and hot years, resulting in annual rises in atmospheric CO_2 concentration [106]. It was most recently suggested that humans may double the number of boreal fires in 'pristine' Siberian forests [107]. If this happened, it would be a departure from the quasi-steady-state, and in the near term, there would be a net flux of carbon to the atmosphere.

Although it seems that fire frequency has increased in many regions, it has decreased in temperate forest biomes as a result of active fire suppression [108]. When active fire suppression excludes fire from the landscape, forests fuels build up and accidental fires become difficult to control, in some cases doing more damage than fires occurring at natural frequencies. This awareness has led to fire policies that include prescribed fires for purposeful regeneration. During extreme dry periods, however, fire in temperate and Mediterranean climates is still a hazard, as was recently the case in California in 2003 and 2007 and in Portugal in 2003 and again in 2005.

Fossil fuel, driving the carbon cycle into the Anthropocene

After food, fossil fuel is today humanity's most needed source of energy. Human activities are emitting 7 ± 0.7 Gt C per year as CO_2; whereas the amount of carbon in food, which provides energy for mankind's 'basic survival,' amounts to a total of 1.2 Gt C per year. In other words, fossil fuel carbon has become a primary fuel for life. Since the beginning of the Anthropocene Era, humans have emitted 300 Gt C of fossil CO_2 and approximately 120 Gt C from deforestation. This represents about twice the increase in atmospheric burden (e.g. [84]), the rest being dispersed by biogeochemical processes into the marine and terrestrial carbon reservoirs. Compared with the 7 Gt C of 'geological' fossil CO_2 emitted yearly, the weathering of silicate rocks removes only 0.07 Gt C per year. This gives a perspective for the extraordinary disequilibrium of the geological carbon cycle that has been caused by humans.

Fossil fuel emissions have increased rapidly in the past decades (see again Figure 8.4). The fossil CO_2 emission curve can be fitted with an increasingly exponential curve, with a characteristic timescale of 37 years. In 1900, the annual coal global mining rate was 100 million times higher than in 1750, when it began. Oil represented only 10% of the fossil fuel in the 1920s, but because of its efficiency and utility for transportation systems, its consumption rose up sharply after

World War II. By the mid 1960s, emission of CO_2 from oil burning exceeded those from coal, and today oil burning contributes approximately 40% of the total fossil fuel CO_2 flux. Finally, natural gas emissions have increased alongside those of oil.

Yesterday, today, and tomorrow's 'Who's Who' in the fossil fuel emissions

Fossil fuel and cement emissions estimates are based on energy statistics. The accuracy of fossil fuel emission estimates is high, on the order of 5% globally. The fossil fuel CO_2 emissions went up from 5.4 ± 0.3 in the 1980s, to 6.3 ± 0.4 Gt C per year in the 1990s,[5] to 7 ± 0.7 Gt C per year during 2000–5 (IPCC Fourth Assessment Report estimates). Such global totals, however, do not capture the large unevenness in the 'Who's Who' of fossil CO_2 emitters.

The Kyoto protocol negotiators were well aware of this, and they distinguished between Annex B countries (the 'old' industrialized nations) and non-Annex B countries (all other nations in the developing world). Currently, the Annex B countries emit 3.9 Gt C per year and non-Annex B countries 3.1 Gt C per year. However, in the non-Annex B category, a few 'heavyweight,' fast-growing economies account for more than half of the emissions. The worldwide distribution of emissions per capita is even more heterogeneous. In the United States, it is slightly more than 5.5 Gt C per year per person (more than four times today's global average), and this emission rate has been relatively constant over the past three decades. In France, it is 1.7 Gt C per year per person, also rather stable over the past decade. By comparison, in China, the fossil fuel emissions per capita in 2002 was 0.73 Gt C per year per person, but it has increased by 14% in one decade.[6] Recently, the International Energy Agency, in its World Energy Outlook 2006,[7] estimated that China will surpass the United States in carbon dioxide emissions before 2010, about a decade ahead of previous predictions. In other words, the 'Who's Who' in fossil fuel CO_2 emissions is changing, and this will need to be recognized in future discussions of the Kyoto protocol.

How much fossil carbon is out there?

The reserves of fossil fuel are estimated to be of 250 Gt C for oil, 200 Gt C for natural gas, and about 4000 Gt C for coal. Needless to say, the geographic and geopolitical distributions of these reserves are uneven. Generally, regions rich in oil and natural gas are far from regions where these fossil fuels are used. For coal, it is the contrary, in particular with China and India, two superpowers of the coming decades that both have enough coal to meet the projected requirements

---

[5] See http://cdiac.esd.ornl.gov/trends/emis/em_cont.html.
[6] All data in http://cdiac.esd.ornl.gov/trends/emis/prc.htm.
[7] See www.iea.org/w/bookshop/add.aspx?id=279.

for their very rapidly expanding economies. The United States, Germany, and Russia also have very large coal reserves. The limit on coal use thus will not be imposed by its physical availability but rather because of its impact on climate and other environmental concerns, such as its detrimental impact on air quality.

The amount of oil reserves and resources has generated controversy between the 'pessimists' and the 'optimists' during the 1990s. The 'pessimists' believe that all of the oil-bearing regions worth exploring have already been explored and that the big fields have already been discovered; thus, future discoveries will be small. They base their conclusions on the statistical study of past discoveries, considering that all oil and gas fields are static objects (with no evolution in the size of initially recoverable reserves). On the other hand, the 'optimists' hold a dynamic concept of reserves and believe that a method based solely on past statistics of discoveries only yields a partial image of the actual potential. The volumes of exploitable oil and gas are closely correlated with technological advances and technical costs, conditioned by the market price of oil and gas. According to an optimist, any improvement in the recovery rate of oil and gas will enable the industry to tap substantial additional reserves. Given the challenge of CO_2 and climate change, the definitions of 'pessimist' and 'optimist' may change.

Similarly, the boundary between conventional and nonconventional hydrocarbons is not fixed, and it could be shifted over time. The unconventional fossil fuels of various types are estimated to be in the range of 15 000 to 40 000 Gt C (of a total sedimentary carbon stock on the order of 1 000 000 Gt C). Among these, the 'oil-like' unconventional fuels include 'heavy oils' (approximately 180 Gt C) that can be pumped like oil but would be more difficult and expensive to refine (e.g. the Venezuela Orinoco Basin) and tar sands (approximately 273 Gt C) that could be mined but at an expensive cost (e.g. Canada's Athabasca tar sands). The Greenpeace report on Climate Change[8] reports that '3000 billion barrels reserve (note = 455 Gt C) of nonconventional oil (out of an ultimate resource base more than an order of magnitude bigger) could then sustain a continued increase in world oil use beyond the middle of the twenty-first century.'

Finally, there are CH_4 clathrates in the seafloor, which have a high-energy content. From the first accepted estimate of stock of 10 000 Gt C [109], the amount has been revised to lower values, now closer to 500–2500 Gt C [110]. We do not know enough about the ways in which the gas hydrates occur underneath the seafloor to be able to plan for their recovery, because hydrates have been drilled in experimental projects only at very few places. If gas hydrates are mainly thinly dispersed in the sediments, they will be very difficult to exploit.

---

[8] See http://archive.greenpeace.org/climate/arctic99/reports/odell317.html.

In summary, humanity will likely not starve fossil fuel during the next 100 plus years. Rather, the risk of a dangerous CO_2-induced threat to the climate system will likely drive 'optimists' and 'pessimists' towards a lower and more realistic 'allowable' level of usable fossil fuel.

The day of tomorrow

In the coming decades, continuing fossil fuel emissions will continue to drive up atmospheric CO_2. The intensity of emissions, and thus the atmospheric load, will depend primarily on economic pathways of energy use and secondarily, but importantly, on how the carbon cycle responds. Economic uncertainties in future emissions are addressed by considering ranges of economic-energy scenarios. The IPCC, for instance, established a set of scenarios with contrasting future development and emission pathways. The way followed by the IPCC was to: 1. develop emission scenarios up to 2100 [3]; 2. calculate the atmospheric CO_2 changes following each scenario; 3. calculate the changes in climate due the radiative forcing changes implied by trends in CO_2; and 4. estimate the diverse impacts of climate changes. An analysis of the economic costs of climate change impacts and of mitigation strategies that could slow down climate change and eventually stabilize the Earth's climate was also made.

Four contrasted families of future scenarios of fossil fuel and land-use CO_2 emissions were designed. The differences in CO_2 anthropogenic emissions among the scenarios are not evident until 2020, a horizon for which predicted emissions are in the range 10–15 Gt C per year. The emission scenarios diverge more, however, after 2050, with some environmentally friendly ones having a decrease in emissions down to today's value in 2100, while others have a continuing rise in emissions, up to about 30 Gt C per year in 2100. Therefore, if the range between the different emission scenarios is a reasonable measure of the 'economic uncertainty,' it is fair to say that the range of possible futures is very large, and that the future emissions pathway will drive the future atmospheric CO_2 concentrations. However, how the natural carbon reservoirs will adjust to rising CO_2 and climate change for each given emission pathway will also be important in controlling CO_2 in the future. For instance, the most recent model simulations of atmospheric CO_2 in 2100, using the 'best available' models, show a spread of 300 ppm (600 Gt C) and about 2.5 °C between carbon-climate models (Figure 8.10). This rough but available measure of the 'natural carbon cycle uncertainty' of 600 Gt C is equivalent to the difference in the integrated fossil emission between extreme economic scenarios [3]. Thus, knowledge of the natural carbon cycle does matter. It matters just as much as knowledge of clean technologies that will help make emission reductions possible.

The changing global carbon cycle: from the holocene to the anthropocene 139

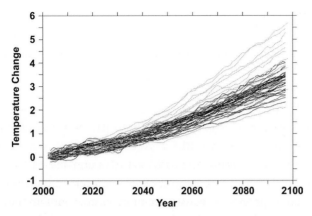

Figure 8.10 Projections of atmospheric temperature variation between 2000 and 2100 from various scenarios with fixed (black) and climate-coupled (grey) carbon dioxide concentrations.

Being able to couple carbon dynamics and climate models is important because it is 'tomorrow's' carbon cycle interacting with 'tomorrow's' climate that must be explored. This exploration is not purely because of scientific interest; there are very real practical issues. Today, nature provides us with an extraordinary 'carbon subsidy' by taking up roughly 50% of the anthropogenic CO_2 loading, but how will it behave in the future? Will the load of anthropogenic CO_2 continue to be improved by the terrestrial 'green sponges' and the oceanic 'carbon pumps,' or will these sink processes weaken? The answer depends on identifying, understanding, and being able to predict both the future of the contemporary and future carbon sink mechanisms. This is one of the scientific 'grand challenges' – it is certainly highly important for society and scientifically difficult given the extraordinarily complex set of interactions involved: ecological, physiological, chemical, and physical, not to mention economic and social. Today's uncertainties in the future uptake of CO_2 by land and oceans are quite large and reflect the scope of the carbon challenge.

On land: green-up or brown-down? Even a small change in the delicate balance between the incoming and outgoing 'gross fluxes' of CO_2 over land can strongly modify the carbon budget of ecosystems. Clearly, for the terrestrial biota to be a net sink of CO_2, the photosynthetic uptake of carbon by plants must be greater than the return flux via respiration and fires. This appears to be exactly what is currently happening [66]. From Earth-observing satellites, there are even hints that photosynthetic activity is increasing – the planet is becoming 'greener' [111]. However, given the relatively small size of the active terrestrial carbon reservoir compared with the ocean (see again Figure 8.7) and the relatively short lifetime (roughly 40 years averaged globally) of carbon on land compared with the ocean, it is most unlikely that this land sink will be maintained indefinitely. However, there will be strong regional differences in the future evolution of the land carbon sinks, and these differences could be important from a carbon management viewpoint.

In projections of future land carbon cycling [112, 113, 114, 115], in which only the beneficial impact of high CO_2 on photosynthesis (the fertilization effect) is included, the land biosphere behaves like a 'green sponge' and soaks up large amounts of carbon from the atmosphere. However, this almost certainly paints an overly optimistic picture. Although fertilization studies, such as forests growing in enriched CO_2 atmospheres, show significant increases in growth [116], there are very real limits. First, plants need an additional supply of nitrogen to benefit from the manna of atmospheric carbon, and there just may not be enough nitrogen available to them (e.g. [117]). Second, the current sink may simply be the effect of regrowing forests and may not continue or respond positively to the increase in atmospheric CO_2. Third, the impact of climate change itself could turn sinks into weaker sinks, if not into sources. Warmer temperatures increase the decomposition of carbon in soils, which may or may not be compensated for by increased growth. The carbon stored in permafrost, some of which has been preserved since the last Ice Age in Northern Russia and Canada, is particularly problematic. As the climate becomes warmer in polar regions, melting permafrost could release large amounts of both CO_2 and CH_4 into the atmosphere. Therefore, again, there is a delicate balance between ingoing and outgoing carbon: both fluxes can be altered by concurrent effects of warming, and this becomes even more complex when one adds considerations of changes to the hydrological cycle.

Changes in rainfall are expected to play a key role in the future response of photosynthesis to climate change. More than 40% of the globe photosynthesis is water limited [118], and this limitation could become more widespread or more stringent in the future [119, 120]. If this happens, the land carbon sink will decrease. A review of a wide range of terrestrial carbon models used to assess responses to the effects of both increasing CO_2 and climate change suggests that future climate *and* CO_2 changes will have 'on average' a detrimental impact on the uptake of anthropogenic CO_2 by land ecosystems [120, 121, 122]. Tropical forests may be especially vulnerable to water stress [119, 123]. In agreement with this finding, new coupled model simulations [124] that integrate ocean, land carbon, and climate models showed that the detrimental effect of climate change alone on the carbon sink was in the range of -2 to -10 Gt C per year. In all of these studies, the negative impact of climate change more than offsets the positive one due to CO_2 fertilization. Yet the response of sinks to climate change is more uncertain than the response to CO_2 changes (the fertilization effect).

In the ocean: physics with the help of algae. In most ocean models (most of them not including marine biology changes), the ocean remains a sink of atmospheric CO_2 in the future. This sink increases with increasing CO_2 in the atmosphere. However, ocean models diverge in their predicted rate of uptake. A region for which the difference among models is the largest is the southern oceans, because the ocean circulation there is poorly known. Adding a further complication is the fact that as anthropogenic CO_2 dissolves into the ocean, the ocean sink tends to decrease in relation to the CO_2 fossil flux because the growth in dissolved inorganic carbon induces an increase in ocean acidity, thereby hampering further uptake.

Climate change also can impact the ocean sink of CO_2 in several ways. First, in response to warming, the CO_2 solubility decreases, and this causes a reduction in the ocean sink. This is a very certain process, already at work today. An additional process by which the ocean uptake could be slowed down by climate change is reduced physical mixing. The surface ocean represents only 2% of the global ocean stock and has a limited uptake capacity (see again Figures 8.7 and 8.8). The transport of carbon between surface and deeper waters by mixing is critical to let more carbon enter into the sea. All ocean models show hints of an increased stratification in the future, which (ignoring possible biological changes) would further decrease the ocean sink.

A second category of feedbacks that are more uncertain concern changes in the marine biota. In the ocean, the export of dead organic matter from the surface to depth drives a burial of carbon from the atmosphere by algae. If this flux of export production (see again the discussion on the ocean biological pump in section "The ocean system: A big carbon processing machine" above) increases (decreases), the ocean will be a greater (smaller) sink due to biology. There are currently no data indicating that the ocean export productivity is changing at a large scale. However, ocean models fitted with a diverse array (from simple to complex) of biological models indicate that changes in ocean circulation caused by climate change could modify the export production and feedback on atmospheric CO_2 and climate. Under future warming scenarios, increased ocean stratification will likely change the carbon dynamics of algae. Decreases in ocean surface mixing could increase the efficiency with which phytoplankton use nutrients, and this would produce an ocean with higher new primary production and, hence, an increase in carbon export. On the other hand, a slowdown of the ocean circulation could also decrease the export production if the supply of nutrients does not satisfy the plankton's demand. Because of this effect, ocean models predict a reduction in the export productivity of a few percent, essentially at low latitudes [125].

Note that any changes in biological processes (plankton growth, particle sinking, mineralization processes) exporting carbon out of the surface will only augment the carbon sequestration during a *transient* phase, when the downward-going export flux of carbon increases. A return loop bringing carbon back up to the sea surface will eventually adjust to the changed export flux, and the net effect of the ocean biota on the atmosphere will be neutral again.

Future changes in ocean biota have the potential to cause some feedbacks on CO_2 and climate change. Currently, there are only few model studies that account for the response of the ocean carbon cycle and the marine biological pumps to climate and atmospheric CO_2, and these studies suggest a rather modest positive feedback [83, 126, 127], on the order of 20 ppm. Furthermore, this biological feedback would develop on longer timescales than the 'immediate' physical and chemical feedbacks. We note again, however, that the subject is complex and not yet adequately considered.

The carbon cycle strikes back. As seen in Figure 8.3, the paleorecord contains strong evidence that the carbon cycle and the climate system are intimately coupled; there is a remarkable and strong correlation between temperature and CO_2 and CH_4 atmospheric concentrations [128]. Over the last decade, the two-way coupling of carbon and climate has been more extensively incorporated into global models revealing new prognostic insights and challenges (see again Figure 8.10). For instance, two studies [119, 123] coupling carbon and climate showed a positive feedback on the carbon cycle of climate change. In the Cox et al. study [119], the future climate looks like what might be termed a 'permanent El Niño' in that dryness extends over the Amazon. The rain forest could not survive in this condition, and this would cause huge losses of CO_2 to the atmosphere. The magnitude of this additional CO_2 load caused by climate change is important compared with the prime impact of the fossil emission scenario (it amplifies the future CO_2 levels by about 20%).

The discovery of an amplification of future CO_2 by climate change has tempered the initial 'nothing to worry about' optimism about green sponges absorbing carbon. Almost all models, which couple carbon and climate, confirm a positive feedback of the carbon cycle, but there is a very large spread between these calculations. Much work remains to be done to narrow down the uncertainties, in particular by validating models with current observations.

The day after tomorrow

Contrary to what is commonly believed when using a 'lifetime' for CO_2 [2], there is no single atmospheric lifetime for removing the excess slug of fossil CO_2. This is because several processes come into play in absorbing atmospheric CO_2 in the immediate, mid-, and long-term future.

During the first few centuries, the main cleaning process will be uptake by the ocean, with uncertainties described earlier. The typical timescales for the entire 'ocean conveyor belt' to adjust to the atmospheric load is on the order of 1000-plus years. After that time, around 15% of the total slug of fossil carbon burned will remain in the atmosphere. The second cleaning process is the equilibration of carbonates. As more and more fossil CO_2 dissolves into the ocean, the ocean acidity increases, and CO_2 is slowly removed by reaction with carbonate sediments on the seafloor [85]. This process is very slow (10 000-plus years). Thus, on the day 'after tomorrow,' the carbonates will be the 500-pound gorilla to neutralize fossil CO_2. There is enough carbonate in the seafloor to neutralize all the fossil carbon. Finally, the geologic 'terminator' that will erase the last traces of the fossil slug is the weathering of silicate rocks. This is an extremely slow process, however, acting on a timescale of over 30 000 years [85].

It is fair to say that uncertainties and feedbacks in the 'after tomorrow' cleanup of anthropogenic CO_2 are large. The ocean warming, for instance, may leave more CO_2 in the atmosphere. For instance, increases in meltwater delivery to the ocean may slow down mixing and therefore the ocean's capacity to sequester atmospheric CO_2. Despite these uncertainties, we can be assured that the fossil carbon legacy of the Anthropocene Era will remain in the atmosphere for a very long time, implying that our descendants will have no choice but to live on a CO_2-rich planet, with dramatically altered climate and ecosystem conditions, unless we act soon.

Stabilizing future CO_2 and climate

What will the long-term fate be of fossil CO_2 in the atmosphere? When will humankind stop emitting fossil carbon? Will it be because of adverse effects of climate change – if so, then it will be *too late*. We are neither proponents of 'A Day After Tomorrow'[9] nor of 'A State of Fear,'[10] but rather we believe that there is an important emerging consensus reflected in the scientific literature and articulated in the Third Assessment Report (TAR)[11] of the Intergovernmental Panel on Climate Change and reinforced in the fourth one (AR4, see Chapter 1):

1. Global mean surface temperature has increased more than 0.5 °C (0.9 °F) since the beginning of the twentieth century, with this warming likely

---

[9] See www.thedayaftertomorrow.com/.
[10] See www.michaelcrichton.net/fear/.
[11] We also know that this consensus is growing and this growth in scientific consensus will be reflected in the Fourth Assessment Report of the Intergovernmental Panel on Climate Change, which is in final review as this paper is being written.

the largest during any century over the past 1000 years for the Northern Hemisphere.
2. An increasing body of observations of climatic and other changes in physical and ecological systems gives a collective picture of a warming world.
3. There is new and stronger evidence that most of the warming observed over the last 50 years is attributable to human activities.
4. Global temperature will rise from 1.4–5.8 °C (2.5–10.4 °F) over this century unless greenhouse gas emissions are greatly reduced.

Yet we also are concerned that even this scientific and policy consensus is, in fact, 'shortsighted.' What is not adequately recognized is the longer-term reality of an inherited and far greater, irreversible disruption of the carbon cycle and induced climate change, even under the most optimistic assumptions about stabilization of the concentration of greenhouse gases. For instance, even the optimistic assumption that the atmospheric concentration stabilizes at roughly twice the preindustrial concentration by the year 2100 (which requires the phasing out of nearly all fossil fuel burning well before 2100) leads to temperature increases by 2100 in the range of 1.4–3.1 °C (2.5–5.6 °F). Furthermore, as we have discussed previously, our current understanding of the carbon cycle suggests that the greater the warming, the larger the positive feedbacks. Hence, even drastic cuts in fossil fuel may not be sufficient.

Also, what is not well recognized is that temperature will continue to increase globally for another 300–500 years *until climate reaches a new stable climate equilibrium with the stabilized value of carbon dioxide* – because of the complex feedback loops involved, climate lags the greenhouse forcing (the effects lag the cause; see Figure 8.11).

Two facts must be stressed:

1. To stabilize the concentration of carbon dioxide in the atmosphere, we must drastically cut (by more than 80%) the global consumption of fossil fuels as well as manage the biosphere far better. In other words, stabilizing the atmospheric concentration of CO_2 requires far more than simply stabilizing emissions – the atmospheric concentration continues to increase almost indefinitely with stable emissions.
2. To stabilize the global mean temperature below levels of harm requires that we acknowledge that global temperature increases lag the increases in the atmospheric concentration of carbon dioxide. Consequently, temperature continues to increase globally well after the stabilization (in the atmosphere) of human-induced greenhouse gases (e.g. principally carbon dioxide).

The changing global carbon cycle: from the holocene to the anthropocene 145

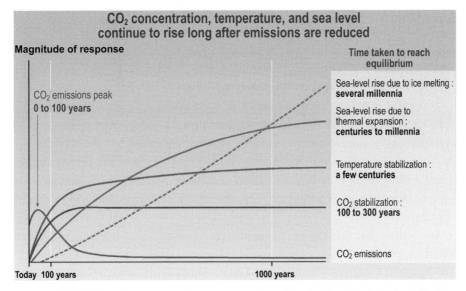

Figure 8.11 CO_2 concentration, temperature, and sea level continue to rise long after emissions are reduced (after [4]).

Almost 15 years have elapsed since the Climate Convention in 1992, and in the meantime, emissions have increased by 15%, a change of larger magnitude than the one agreed on for Annex B countries in the Kyoto Protocol. Since 1992, atmospheric CO_2 has increased by 25 ppm. This amount of CO_2 carries a warming of 0.25 °C (0.45 °F); we need to alter course.

What should we do to reduce uncertainties?

Earlier, we stated that the geographic distribution of the land and ocean sources and sinks of CO_2 has remained elusive and that this uncertainty plagues future CO_2 projections. Understanding and perhaps managing carbon sinks (and other carbon sources) will be important, as will having independent knowledge of fossil fuel sources.

Since the atmosphere is a fast *but incomplete* mixer of surface CO_2 fluxes, the geographical distribution of CO_2 in the atmosphere can be used to quantify surface fluxes. Most of the current atmospheric measurements (fewer than 100 monitoring sites globally) are marine stations, where sparse sampling may be representative of the CO_2 concentration over very large scales. This sparse set of atmospheric CO_2 measurements has been used with atmospheric circulation/transport models to quantify regional sources/sinks of carbon. These atmospheric data strongly suggest the existence of a significant Northern Hemispheric sink for carbon that is likely to be terrestrial. Unfortunately, estimation of *regional* carbon fluxes is a significantly underdetermined problem. For some regions like North America or the vast expanse of Eurasia, we do not know even if the region is acting as a net carbon source or a net carbon sink from year to year. Simply stated, the current set of direct *in situ* observations is far too sparse, and this sparseness hinders geographically resolved source-sink determination.

Long-term, accurate measurements of atmospheric CO_2 with global 'wall-to-wall' coverage would shift the problem from being an underdetermined or indeterminate system to an overdetermined problem that would be beaten into submission by data. This shift is fundamental; it would allow the determination and localization of CO_2 fluxes in time and space. Measurements that densely sample the atmosphere would capture source/sink regions and provide a crucial constraint.

We believe that this objective is within reach in the next 5–10 years and that it should be the cornerstone of an Integrated Carbon Observing System (Figure 8.12). The Integrated Carbon Observing System should be built on high-precision, long-term, ground-based CO_2 observation networks and on new satellite observations that will fill in the gaps of the ground-based network in regions that are virtually impossible to sample.[12]

What is needed for the ground-based CO_2 network is a global, distributed infrastructure capable of sustaining atmospheric measurements of CO_2 and related tracers at the highest accuracy, with minimal risks of hiatus in data over several years. Components of this future infrastructure already exist in the United States with the NOAA-CMDL global air-sampling program, but in Europe there needs to be a better-integrated and more harmonized observing system. Developing common methodologies, standards, data management systems, protocols, and instrumentation will increase the cost efficiency of the global *in situ* observations by avoiding duplication and by facilitating data sharing.

A first step in measuring CO_2 from space is being made. In 2009, NASA will launch the OCO (Orbital Carbon Observatory) instrument, which will measure the average content in CO_2 of each air column from the attenuation of solar photons traveling across the Earth atmosphere. The Japanese space agency JAXA will also launch the GOSAT mission, developed in cooperation with the National Institute for Environmental Studies.

What is needed for space-borne measurements is a highly precise global data set for atmospheric CO_2 without seasonal, latitudinal, or diurnal bias. It is possible with current technology to acquire this data set with a space-borne sensor using multiple-wavelength laser absorption spectroscopy and mature CW fiber lasers

---

[12] For a more extensive discussion of this system, see http://ioc.unesco.org/igospartners/Carbon.htm.

The changing global carbon cycle: from the holocene to the anthropocene 147

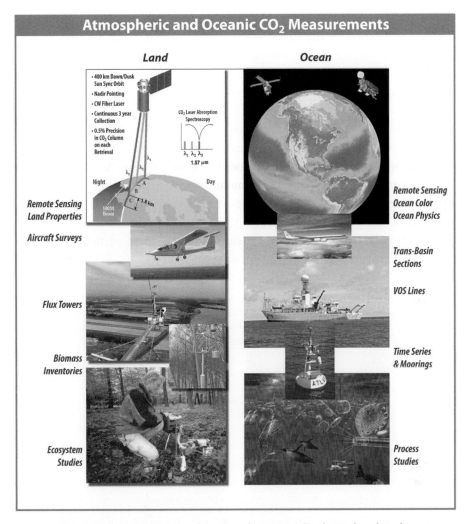

Figure 8.12 An integrated carbon observing system. (For image in color, please see Plate 9.)

developed by the commercial telecom industry. We believe that this mission would ideally be an international mission – perhaps a US-French mission.

The coupling of this high-precision, high-volume data stream combining *in situ* and satellite observations with atmospheric inversion, data assimilation, and coupled atmospheric, terrestrial, and ocean carbon modeling will enable us to quantify the CO_2 sources and sinks at unprecedented space and time resolution. The final scientific outcome will be a greatly advanced understanding of the global carbon cycle as well as of the essential scientific foundation for making improved projections of future atmospheric concentrations of CO_2. Yet even more importantly, we will have put in place one important brick in the infrastructure that will be needed for the next century to address changes in the environment of our planet.

A closing thought – reducing uncertainties is not enough

The linked challenges of confronting and coping with climate and other environmental changes and addressing and securing a sustainable future are daunting and immediate, but they are not insurmountable. The challenges can be met, but only with a new and even more vigorous approach to understanding our changing planet and ourselves and with a *concomitant commitment* by all to alter our actions. Those who consume the most must take the greatest actions. We simply must take some of the pressure off the Earth. In particular, we must begin *now* to reduce fossil fuel emissions.[13]

[13] We acknowledge that we (Moore) expressed similarly this thought in a prior publication: W. Steffen, J. Jäger, D. Carson and C. Bradshaw, eds., *Challenges of a Changing Earth: Proceedings of the Global Change Open Science Conference, Amsterdam, The Netherlands, 10–13 July 2001* (New York/Tokyo/Berlin/Heidelberg: Springer-Verlag, 2002).

9

Challenges of climate change: an Arctic perspective and its implications

ROBERT W. CORELL

> The Arctic Climate Impact assessment is path-breaking and it is crucial that the world know and understand what it says.
>
> –Sheila Watt-Cloutier, Chair of the Inuit Circumpolar Conference, which represents internationally the Inuits who live in Alaska, Canada, Greenland, and Chukotka, Russia

Overview of the Chapter

This chapter addresses four questions as a means of outlining the issues of climate change and then sets them in an Arctic framework, because the region is experiencing the most dramatic current and projected changes in climate. In this context, the Arctic is the planet's bellwether for climate change.

This chapter therefore addresses the following questions:

1. What does science tell us about our changing climate, particularly recent findings that have global implications?
2. What does the Arctic case tell of the character of recent climate change?

Robert W. Corell has been actively engaged in research concerned with both the science of climate and global change and the interface between science and public policy. He was Assistant Director for Geosciences at the National Science Foundation, where he led the US government's research programs on climate and global change (1986–2000), including the US National Assessment of Climate Change. He led the international Arctic Climate Impact Assessment (2000–2005) and was a senior research fellow at the Kennedy School of Government at Harvard University (2000–2004), engaged in research on global environmental issues and sustainable development. He is now Senior Policy Fellow at the Policy Program of the American Meteorological Society, where he continues his research and assessment activities.

3. What are some examples of strategies/scenarios toward emission reduction?
4. What are the consequences and/or risks of inaction?

What does science tell us, particularly recent findings that have global implications?

The foundations of our collective insights into the planet's changing climate have been greatly enhanced by a comprehensive international scientific research effort over the past several decades that has markedly increased our understanding of the character, processes, and effects of climate change. Furthermore, both international and national studies to assess the character and consequences of climate change have more fully documented the impacts of these changes on the planet and its peoples. For example, the scientific evidence documented by the 2001 Intergovernmental Panel on Climate Change (IPCC) assessment [3] concluded with high confidence that '[i]n the light of new evidence and taking into account the remaining uncertainties, most of the observed warming over the last 50 years is likely to have been due to the increase in greenhouse gas concentrations.'

2005 global temperatures are the warmest on record

The 2005 global-mean surface temperature anomaly was the warmest on record, with both land and oceanic temperatures continuing to warm. The significance of the continued increases in oceanic temperatures is that the oceans absorb over 90% of all warming, with only 3.3% going to warm the atmosphere and 6.2% going to melt sea ice and glaciers. The warming of the oceans cannot be explained solely by natural internal climate variability or by solar and volcanic forcing, but the measured temperature increases are well simulated by anthropogenically (i.e. human induced) forced climate models, suggesting that the warming of the world's oceans is of human origin. Furthermore, analyses suggest that Earth is now absorbing 0.85 ± 0.15 W/m^2 more energy from the Sun than it is emitting back into space. This has a profoundly important implication: there is already stored in the oceans an additional global increase in temperature of about 0.5 °C (0.9 °F), even without any further addition of greenhouse gases.

In summary, it is increasingly clear that the planet is warming and has been doing so for much of the past 50 years and that the primary source of that warming has been derived from human activities, particularly the use of fossil fuels, and that the projections for the future are profoundly worrisome and will need to be aggressively addressed by the policy communities at all levels of society. The scientific peer-reviewed evidence underpinning these summary observations is solid, credibly documented, and consistent with virtually every scientific assessment of climate change of recent years.

What does the Arctic case tell us of the global implications of climate change?

Over the past five years, more than 300 scientists and experts, including elders of the indigenous peoples and other insightful residents, have worked on a comprehensive analysis, synthesis, and documentation of the impacts and consequences across the Arctic of climate variability and changes known as the Arctic Climate Impact Assessment (ACIA). Through the Arctic Council, a ministerial meeting of the eight Arctic countries charged a group of international scientists to assess the impact of climate change on the Arctic. A scientific report was consequently produced that contained two documents: a synthesis of the principal findings [129] and a comprehensive scientific analysis of climate change totaling over 1000 pages in 18 chapters [130]. Also, the Arctic Council produced a policy recommendations report [131].

Major findings

The ACIA details and projects significant disruptive impacts from climate change while identifying a number of potential opportunities for indigenous and other residents, communities, economic sectors, and governments of the region. To develop its projections, the assessment used two scenarios from the IPCC Special Report on Emissions Scenarios [35], the B2 and A2 scenarios (see Chapter 1). B2, the primary scenario, is a 'moderate' climate change scenario that projects global carbon dioxide emissions to more than double by 2100. The B2 scenario describes a world in which the emphasis is on local solutions to economic, social, and environmental sustainability, whereas the A2 scenario describes a very heterogeneous world with an underlying theme of self-reliance and preservation of local identities. Under the B2 scenario, the Arctic warming is about twice that projected for the global, with some regions of the Arctic having substantially larger warming possibilities.

Evidence of recent warming in the Arctic includes records of increasing temperatures, melting glaciers, reductions in extent and thickness of sea ice, thawing permafrost, and rising sea level. There are regional variations and patterns within this overall trend; for example, in most places, temperatures in winter are rising more rapidly than in summer. In Alaska and western Canada, average winter temperatures have increased by as much as 3.0–5.0 °C (5.3–9.0 °F) over the past 30 years, while the global average increase over the past 100 years has been only about 0.6 ± 0.2 °C (1.1 ± 0.4 °F).

Over the past 30 years, Arctic sea ice extent in the summer months has decreased on average by about 25%, and this change has been 20% faster over the past two decades than over the prior three decades. The ACIA model simulations for the decades ahead project substantial and accelerating reductions in summer sea ice around the entire Arctic Basin, with one model projecting an ice-free Arctic in the summer by the middle of this century, leading to seasonal opening of potentially important marine transportation routes while producing significant changes in albedo, alterations in cloudiness, and potential changes in ocean circulation that have global importance. The implications for access to the Northern Sea Route along the Eurasian coast from the Atlantic to the Bering Strait is an increase of the period of opening from a few weeks to 100 days or so and to 150 days for moderately capable ice-breaking vessels by 2080. This has the potential for important economic and political implications; for example, increasing access to the region's resources and raising issues of sovereignty, safety, and environmental preservation. Access to sea ice is critical for the survival and reproduction of many high-latitude marine mammals. Scientists and Arctic residents are concerned that the thinning and depletion of sea ice in the Arctic will cause the extinction of key marine mammals, including polar bear, walrus, and some species of seal. Loss of these species threatens the hunting culture of Inuit in Alaska, northern Canada, Greenland, and Chukotka, Russia.

Recent studies of glaciers in Alaska already indicate an accelerated rate of melting, representing about half of the estimated current loss of mass by glaciers worldwide. The documented melting in Greenland is of special importance, because Greenland has the potential to increase sea level substantially. Over the past two decades, the melt area on the Greenland ice sheet has increased on average by about 0.7% per year (or about 17% from 1979–2002), with considerable interannual variation. Satellite data indicate that in 2005, the surface area of melt was the largest on record. Although thermal expansion of oceanic waters from increased temperatures will dominate the increases in sea level, the ACIA assessment suggests that the ice-sheet melting will contribute to an increase in sea level that is likely to be at the upper end of the IPCC projection (i.e. 0.2–0.9 m) during this century. One meter of sea level rise has been estimated by Bangladesh officials (by the minister of environment at a recent UNEP Ministerial Session) to reduce the usable land area of their country by 60% (while other estimates place the reductions in the 25%–40% range). Many of the small island developing states will be inundated by the rising sea levels, and a 1-m rise will reduce the land area of a number of these countries by about 50%.

Permafrost presently underlies most of the land surfaces in the Arctic, and thawing ground will disrupt transportation, buildings, and other infrastructure.

Permafrost temperatures over most of the sub-Arctic land areas have increased by up to 2.0 °C (3.6 °F) over the past few decades, and the depth of the layer that thaws each year is increasing in many areas. Over the next century, permafrost degradation is projected over 10%–20% of the present permafrost area, and the southern limit of permafrost is projected to shift northward in some regions by as much as several hundred km. Rising temperatures are already degrading land routes over frozen tundra and across ice roads and bridges, and the incidence of mud- and rockslides and avalanches is likely to increase in the future. The number of days per year in which heavy equipment can travel on ice roads across the tundra, approved by the Alaska Department of Natural Resources, has dropped about 50% in the past 30 years (from around 200 days per year to around 100 days), limiting oil and gas exploration and extraction and, of course, access by other interests.

Rising temperatures across the Arctic are projected to lead to enhanced growth, denser vegetation, and the expansion of forests into the tundra and from the tundra into polar deserts. This change, along with rising sea levels, is projected to shrink tundra area to its lowest extent in at least the past 21 000 years, potentially reducing the breeding area for many migratory bird species and the grazing areas for animals that depend on tundra and polar desert habitats. Half of the current tundra area is projected to disappear in this century. Inland peoples throughout the Arctic depend on caribou and reindeer herds, which need abundant tundra vegetation and good foraging conditions, especially during the calving season. In addition to reducing the area for grazing, climate-induced changes are projected to increase incidence of freeze-thaw cycles and freezing rain, both of which prevent animals from eating iced-over vegetation. Further, migrations of other species (moose, red deer, etc.) into traditional pastureland are likely to disturb some populations. Although much of the redistribution of species is climate induced, the development of roadways, pipelines, and other infrastructure also contributes. Although Arctic agriculture is a small industry in global terms, the region's potential for crop production is projected to advance northward.

Marine fisheries are a vital part of the economy of virtually every Arctic country and provide an important food source globally. Because they are largely controlled by factors such as local weather conditions, ecosystem dynamics, and management decisions, projecting the impacts of climate change on marine fish stocks is difficult. Based on available information, however, projected warming is likely to improve conditions for some important Arctic fish stocks, such as cod and herring, while negatively affecting others. An example of the latter is the extent of northern shrimp, which will probably contract, reducing the large catch (about

100 000 tons a year) currently taken from Greenlandic waters. Although the total effect of climate change on fisheries will likely be less important than decisions regarding management, specific communities that are heavily dependent on fisheries may be dramatically affected.

Across the Arctic, indigenous peoples accustomed to the wide range of natural climate variations report changes that are unique in the long experience of their peoples [129, 130, 131, 132]. However, the changing climate is altering their ability to project the future. One elder in western Russia made the following observation in an interview, noted frequently by the ACIA across the Arctic basin by many indigenous residents:

> Nowadays snows melt earlier in the springtime. Lakes, rivers, and bogs freeze much later in the autumn. Reindeer herding becomes more difficult as the ice is weak and may give way.... All sorts of unusual events have taken place. Nowadays the winters are much warmer than they used to be. Occasionally during winter time it rains. We never expected this; we could not be ready for this. It is very strange.... The cycle of the yearly calendar has been disturbed greatly and this affects the reindeer herding negatively for sure....

Residents of the Arctic are likely to face major impacts due to climate and other environmental changes, which are occurring in the context of other interrelated changes. Other environmental changes (chemical pollution, habitat destruction, and over-fishing), together with social and economic changes (technological innovations, trade liberalization, urbanization, self-determination movements, and increasing tourism), impact these residents of the north. It appears, though, that the rapid rate of climate changes is limiting their capacities to adapt.

The projections for future climate changes in the Arctic are outside of human experience and are highly likely to impose substantial dislocations, impacts, and vulnerabilities to all residents of the Arctic. Furthermore, because of the direct linkages to planet-wide impacts from sea level rise, the opening of new natural resources and newly available sea routes and changes in global temperature distributions are highly likely to have profound global impacts. The 10 key findings of the ACIA are:

1. Arctic climate is now warming rapidly, and much larger changes are projected (Figure 9.1).
2. Arctic warming and its consequences have worldwide implications;

Challenges of climate change: an Arctic perspective and its implications 155

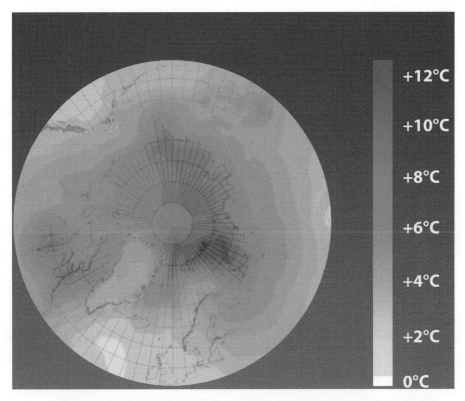

Figure 9.1 ACIA projections of surface temperatures over the twenty-first century based on the average of the five ACIA computer models and using the B2 scenario (source: ACIA Report [129, 130]). (For image in color, please see Plate 10.)

3. Arctic vegetation zones are very likely to shift, causing wide-ranging impacts.
4. Animal species' diversity, ranges, and distribution will change.
5. Many coastal communities and facilities face increasing exposure to storms.
6. Reduced sea ice is very likely to increase marine transport and access to resources.
7. Thawing ground will disrupt transportation, buildings, and other infrastructure.
8. Indigenous communities are facing major economic and cultural impacts.
9. Elevated ultraviolet radiation levels will affect people, plants, and animals.
10. Multiple influences interact to cause impacts to people and ecosystems.

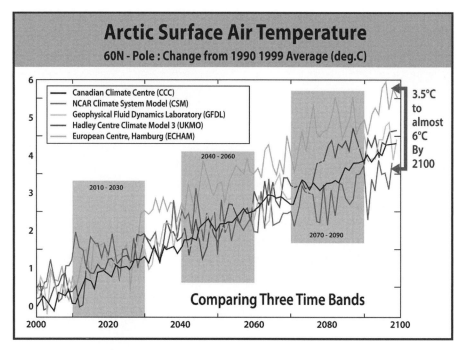

Figure 9.2 Arctic surface air temperatures projected by the five ACIA climate models using the B2 scenario (source: ACIA Report [129, 130]).

What are some examples of strategies/scenarios toward emission reduction?

Strategies and scenarios

Scenarios (see Chapter 1 for an expanded discussion of scenarios) are a plausible set of descriptions of how the future may unfold. Scenarios are neither predictions nor forecasts; they are simply a methodology that is designed to provide a set of plausible descriptions of possible future (e.g. from 2000–2100) states of the world. The scenarios that have been used most extensively during the past few years were developed by the IPCC (Figure 9.2).

International agreements

Although the IPCC scenarios provide a broad range of plausible future states of the planet, they do not provide more detailed information on the pathways and policies of adaptation implementations and mitigation strategies and agreements. Formal international mechanisms do address reductions of greenhouse gases and other mitigation and adaptation measures through the Kyoto Protocol of the United Nations Framework Convention on Climate Change

(UNFCCC). The protocol encourages governments to cooperate with one another, improve energy efficiency, reform the energy and transportation sectors, promote renewable forms of energy, phase out inappropriate fiscal measures and market imperfections, limit methane emissions from waste management and energy systems, and protect forests and other carbon 'sinks.'

Current thoughts and strategies

It is becoming increasingly clear that current measures are not likely to prevent further increases in CO_2 emissions within the timescale of the Kyoto Protocol (2008–12) as a result of current and expected growth of production, consumption, transport, and other factors. To address this, the European Union, for example, declared in the late 1990s that global average temperatures should not exceed 2 °C above preindustrial level. Prime Minister Tony Blair indicated that the United Kingdom should reduce greenhouse gas emissions by 60% by 2050. National governments, state and provincial governments, and even cities have announced similar targets and timetables. For example, the six New England states in the United States and the four Maritime Provinces have committed to participate in the achievement of regional Kyoto-like goals.

What are the consequences and/or risks of inaction?

There is an asymmetry between the timescales for the climate system to react to increases in greenhouse gases and the timescales to recover from such increases. It takes roughly 10 times as long to recover than it takes to increase global temperatures from atmospheric increases in greenhouse gases. There are many factors, but some of the more important ones are:

1. very long residence time of the greenhouse gases in the atmosphere;
2. the ability of the oceans to absorb the heating and transport it around the planet; and
3. the long delays in melting of sea ice– and land-based glaciers.

It is clear from discussions here and in other chapters that CO_2 is the dominant greenhouse gas; however, if methane (CH_4) ever becomes an important greenhouse gas, even with a slight increase in the percentage of emissions, with its global warming potential that is about 22 times that of CO_2, the impact on the warming of the planet will be profound. About 60% of atmospheric methane is generated from human-related activities, according to the IPCC. Methane increases in the past 200 years are due mainly to increased burning of grasslands, forests, and wood fuels; use of landfills; more intense livestock activity and other agricultural activities; coal mining; wastewater treatment; rice cultivation; and natural gas leaked from fossil fuel production. Natural sources of methane include wetlands, termites, oceans, hydrates, and wildfires. For example, it has been estimated that the West Siberian (Russia) bog alone contains some 70 billion tons of methane, a quarter of all the methane stored on the land surface worldwide. Recent assessments suggest that the potential for releases of methane during the coming century are real, although the timescales are uncertain. However, if methane releases begins at some threshold of increased temperature, the only way to halt their continued release is to lower the global temperatures, which is unlikely. The release of methane creates a positive feedback process, thereby further enhancing the effect of methane as a greenhouse gas and hence leading to increases in global temperatures.

The simple message, based on the IPCC and ACIA assessments, is that delays at the beginning (now) only create substantially longer delays in the out years. Given these realities, it is clearly essential that societies implement both adaptation and mitigation measures on timescales that will limit both the magnitude and timing of climate change. Delays in such action have the potential of prolonging both the effects and magnitudes of the consequences of climate change. Therefore, both adaptation and mitigation action plans should have both near- and long-term strategies and implementation elements.

10

On some impacts of climate change over Europe and the Atlantic

JEAN-CLAUDE ANDRÉ

Beaucoup d'eau du Rhône s'écoule pendant tout l'hiver, une raison pour laquelle les hydrologues étudient les inondations.

– J. Martin, 2003

Heat waves

Introduction

France and, more generally, Western Europe experienced exceptionally high temperatures during the summer of 2003. Both the length and spatial extension of this 2003 climatic phenomenon had never been observed over the last 50 years: the most recent 'warm' summers (1976, 1983, and 1994) did indeed affect only a smaller part of this region (Figure 10.1). One of the issues is of course to know whether this phenomenon was the result of the anthropogenic greenhouse warming or whether it was only a 'natural' extreme event. What we call climate is indeed characterized by the mean of a number of events and not by a single one, however exceptional it may be. Whatever its origin, either natural or anthropogenic, the consequences of such an event on human health were extremely large, with a death toll of about 15 000 people in France only, and a total of approximately two times more for all Western Europe. It is then of utmost importance

> **Jean-Claude André** has been working mainly on fluid turbulence, geophysical boundary layers, atmospheric and climate dynamics, and their numerical simulation. He was head of research at Météo-France up to 1994 and, since then, has been the Director of CERFACS (European Centre for Research and Advanced Training in Scientific Computing). He is a corresponding member of the French Academy of Sciences and founding member of the French Academy of Technology.

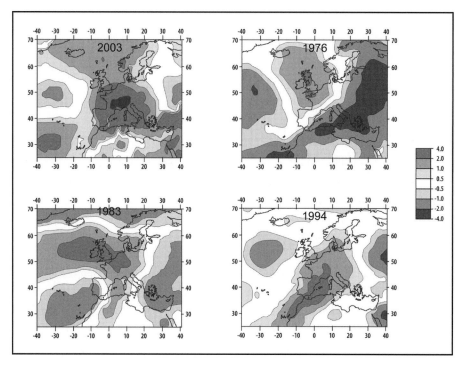

Figure 10.1 Temperature anomaly (in °C) at 850 hPa during June–July–August 2003 (upper left), June–July–August 1976 (upper right), June–July–August 1983 (lower left), and June–July–August 1994 (lower right). (For image in color, please see Plate 11.)

to study if such phenomena are likely to increase in frequency and/or intensity in the forthcoming years as a result of anthropogenic greenhouse warming.

In order to study how the frequency and intensity of rare events like these heat waves might evolve in the future, one needs, of course, to have very long and homogeneous time series. The rarer the event, the longer the series must be. From a statistical standpoint, the correct characterization of an event that takes place every five years on average requires a 100-year-long stationary series. Without such a long and stationary time series, it is impossible to study rare and extreme events from past meteorological records. Because we already know that climate has experienced some changes over the past century (i.e. a slow but constant evolution, with an increase of 0.6 °C [1.1 °F] of the global mean temperature; see Chapter 1), we have to turn to other ways of characterizing and predicting future changes in heat wave frequency and intensity.

Numerical simulation of climate

The only solution to the above problem is numerical simulation of present (and future) climate. One can then be sure that the data time series are

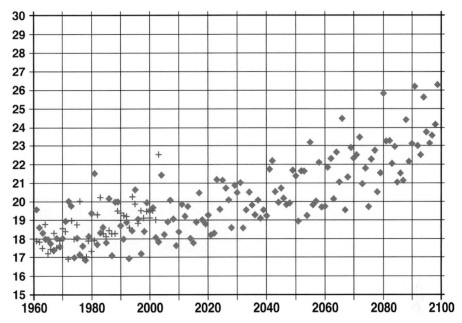

Figure 10.2 Mean summer temperature (°C) over France from 1960–2003, as observed (crosses) and simulated (diamonds) using the 'Arpège-Climat' model (after M. Déqué).

homogeneous, and one can extend the length of the simulations or perform many simulations under slightly different conditions (concerning either initial and/or forcing conditions, or the description of some physical processes) in order to reach statistical significance. The issue here is not so much to look for the best physical climate model, but to achieve statistical significance by adequately sampling the internal variability of the model. Of course, the amount of available computing power then becomes the limiting factor, but technical progress has been providing steadily more computing power over the past decades, and this trend is likely to continue over the next few decades. One could also run the numerical model with a much stronger forcing, e.g. doubling the carbon dioxide atmospheric concentration over a short time, which leads to an increase of the anthropogenic-signal to natural-noise ratio. However, this has one major drawback: the model series might not exhibit the same features as climate natural variations, particularly with respect to very large fluctuations (climate modelers indeed know from experience that fluctuations in model climates are usually somehow smaller than real climate fluctuations).

To illustrate the type of results one can achieve, temperature time series simulated by the 'Arpège-Climat' model of Météo-France are presented here (Figure 10.2). Although the model is global, it includes a variable spatial resolution so that Western Europe is described at a 60-km resolution, which is small enough to capture many processes influencing local climate. Three different simulations were used, each one covering the last 40-year period, in order to compute the distribution of temperature and of other meteorological parameters with good statistical and spatial accuracies.

The comparison of these simulation results with temperature observations performed over the past 50 years shows that the model is reasonably accurate as far as mean values are concerned. It also describes relatively faithfully the periods of high temperature during summer and of high precipitation during winter. It however fails to reproduce cold episodes in winter as well as the high-precipitation events that might occur during summertime. Such biases can nevertheless be corrected by some ad hoc techniques using the relative values of fluctuations rather than their absolute values. Once the biases are corrected, one can deal with model series, which, by construction, exhibit the same statistical properties as the real-world series.

The future of heat waves

If the above simulations are performed over a century, i.e. until the end of the twenty-first century, with for instance a prescribed atmospheric concentration following the A2 scenario defined by the Intergovernmental Panel on Climate Change (IPCC) (see Chapter 1), one can then analyze the 140-year series of the daily model data. Figure 10.2 presents the evolution of the mean temperature over France for the entire period and shows a clear tendency for the mean temperature to increase steadily. One should emphasize that, although individual model years do not correspond to real calendar years, they are possible examples of what could happen, and only the mean features they exhibit are significant. In particular, the model year 2003 shown here does not exhibit the same extreme value as observed (see blue crosses on the figure). Had this happened, it just would have been the result of chance. An additional remark concerns the large interannual (year-to-year) fluctuations that is superimposed over the mean warming until c. 2040. By solely looking at data until 2040, it would be difficult to predict that the mean temperature would increase so much toward the end of the century. This is one of the reasons why scenario studies mostly concentrate on the later period, i.e. from 2070–100. It further justifies the use of long homogeneous model series instead of shorter and nonstationary observational series. Extremes in the model series show that the warmer summers of period from 1961–2000, either observed or simulated, are only marginally warmer than the colder summers of the late century. Furthermore, one does not find any mean temperature lower than 18 °C (64 °F) after 2040, while temperatures higher than 24 °C (75 °F) are found after 2065. This striking result is very robust with respect to the climate model used, as

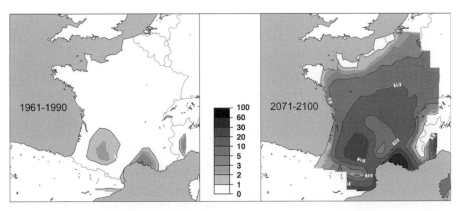

Figure 10.3 Number of days when maximum temperature is above 30 °C during at least 10 consecutive days, for 1961–90 (left) and 2071–100 (right) (after M. Déqué).

the same type of study performed with other climate models (like the one from IPSL) does exhibit very similar features.

Using the same 'Arpège-Climat' model and looking at the frequency distribution of extreme temperature and performing the appropriate statistical analysis, it is possible to characterize the number of days that might be warmer than a given threshold during future summers. Figure 10.3 shows the mean number of days when maximum temperature exceeds 35 °C (94 °F). Nowadays, only a small region in southeast France experiences one or two such episodes each year. Toward the end of the century, the number of such episodes is expected to reach 20 days per year, while for most other French regions, five very hot spells might happen each year.

Such an increase in the frequency and intensity of summer heat waves is not directly linked to variations in water resources. It may nevertheless lead to decreased water quality by allowing proliferation of algae or other microorganisms. It may also lead to increased needs for water, either industrial water (e.g. cooling of power plants) or agricultural water (e.g. irrigation).

Hydrological cycle

Precipitation

Results from 19 different models have been recently compared [133] in the case of a doubled atmospheric carbon dioxide (CO_2) concentration: all models show an increase of mean precipitation almost everywhere (equatorial and mid- and high-latitude regions), except for some decrease in the subtropical belt. The magnitude of this increase might reach locally 5% to 15% or even more, with a mean value of about 2.5%, equivalent to 0.07 mm per day. This mean value varies from one individual model to another (between −0.2% and 5.6%). Although such precipitation changes would impact many human activities, they nevertheless remain of moderate amplitude.

Doubling atmospheric CO_2 concentration also leads to an increase in simulated variability and amplified extreme events. The standard deviation of precipitation increases by 4%, i.e. 0.04 mm per day, with values in individual models ranging from 3% to 11%. Only one model simulates a decrease in standard deviation. Here also, regional variations can reach larger values, such as 15% for instance, in the tropical and high-latitude regions.

To illustrate these results, we present Figures 10.4 and 10.5, showing, for a particular model, the change in precipitation as simulated for a B2 scenario (see Chapter 1) between the mid twentieth century and the mid twenty-first century. Precipitation increases during boreal winters (December to March) over mid- and high latitudes of the Northern Hemisphere and over the ITCZ (Inter-Tropical Convergence Zone), but it decreases in subtropical regions on both sides of the ITCZ. The response is quite symmetric during boreal summer (June to September) in the tropics, with also a northern migration of the ITCZ, an increase in monsoon rains over West Africa and South Asia, and a relative drying of mid-latitudes in the Northern Hemisphere, while high latitudes usually remain wetter compared with the present climate.

In general, however, climate models are unable to reproduce the present location of the convergence zones that experience intense precipitation and of subsidence regions such as over subtropical deserts. Variations of about 10° might appear among models when looking at the latitudes where either the maximum or the minimum of precipitation take place. This implies that it is very difficult to predict the real impact of climate change on water resources in many subtropical regions, such as Africa. Changes in African water resources are nevertheless of crucial importance, as the two most devastating events since 1974 in terms of human deaths are attributed to droughts (Ethiopia-Sudan, 1984, 450 000 deaths; Sahel, 1974–5, 325 000 deaths), exceeding the death toll due to the Indian Ocean tsunami of December 2004.

Impacts on soil moisture

Another impact of climate change is indeed related to the possible increase in droughts that might affect some regions, because such events have a very profound influence on living and economic conditions. Long-enough model runs (300 years) have shown that soil moisture increases in wintertime at mid- and high latitudes but might decrease significantly during summertime: summer

On some impacts of climate change over Europe and the Atlantic   165

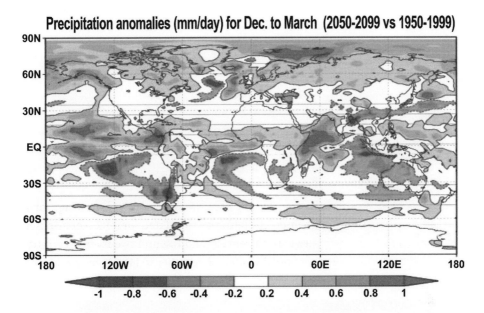

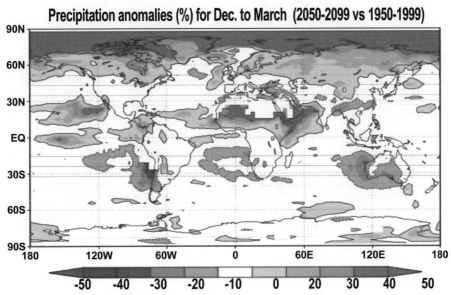

Figure 10.4 Global precipitation anomalies for the twenty-first century (top: mm/day; bottom: %) as simulated for the December–March period in the case of a B2 scenario with 'Arpège-Climat' model (after M. Déqué). (For image in color, please see Plate 12.)

drying might occur over Europe, while winter anomalies exhibit a north-south gradient, with wetter soil at high latitude and dryer soils around the Mediterranean. In some cases, such summer drying might be so intense that water resources could remain scarce throughout the entire year.

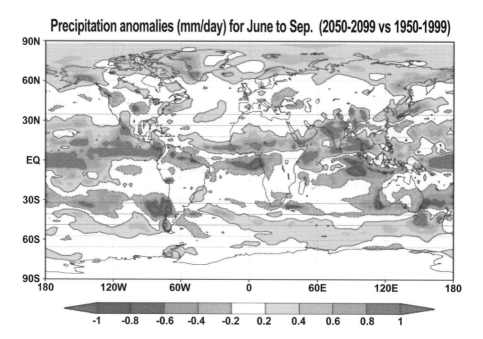
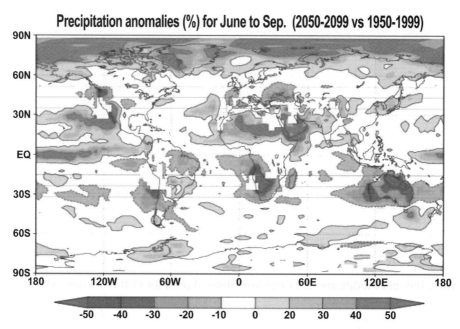

Figure 10.5 Global precipitation anomalies for the twenty-first century (top: mm/day; bottom: %) as simulated for the June–September period in the case of a B2 scenario with 'Arpège-Climat' model (after M. Déqué). (For image in color, please see Plate 13.)

The predictions of soil moisture anomalies are however quite uncertain, as no measurements presently allow the establishment of a climatology against which the model could be validated and calibrated.[1] Simulations nevertheless indicate that soil moisture variations could be detected unambiguously toward the midst of the twenty-first century, i.e. slightly later than the temperature change (see above).

Impacts on snow cover

Global scenarios indicate a progressive retreat of snow cover in the Northern Hemisphere, a feature that has already been observed from space over the period from 1979–99 (a decrease of about 60 000 km^2 per year during wintertime – see Chapter 7). Such a decrease is predicted to get larger during the twenty-first century and could reach a value of about 10^7 km^2 in winter, i.e. roughly 20% of the actual surface presently covered by winter snow. This is important, because the snow cover is very efficient at storing water for later spring and summer release (see Chapter 5). Such snow retreat will also significantly impact the development of tourism during wintertime.

Mountain snow cover

Climate warming plays a double role: on the one hand, precipitation will occur more frequently in the liquid phase (i.e. rain), and on the other hand, snowmelt will start earlier in the season. Such effects are to be felt particularly in mid-altitude mountainous regions, where temperature conditions are close to those of melting. Impact studies have indeed shown that one could expect a mean reduction by two weeks in the snow season above 3000 meters and by up to one month at 1500 meters.

Snow melt

Similar studies have been conducted of the impacts of snowmelt timing on French rivers, especially on those being fed for a significant fraction by snowmelt. Results indicate that future river runoff varies slightly from one model to another, with some climate change features leading to systematic effects: the large sensitivity of the snow cover, both in the Alps and in the Pyrenees, induces modifications of the hydrological regimes of these rivers, with for instance an increase in winter flow (due to both fewer snow-covered surfaces and increased liquid precipitation) and the occurrence of earlier spring floods (see Chapter 5).

---

[1] This situation might change as satellite measurements may become available in the coming years, with the launch in 2008 of the European Space Agency's SMOS mission.

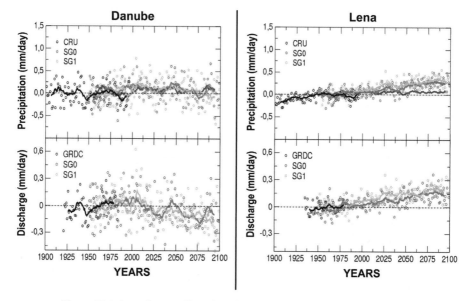

Figure 10.6 Annual anomalies (circles) and 11-year moving average (curves) for precipitation (top) and discharge (bottom), for both observations (black symbols) and simulation (red symbols), in the case of the Danube (left) and the Lena (right) rivers. (For image in color, please see Plate 14.)

Impact on river runoff

Large-scale river runoff models usually have a spatial resolution of about 1° × 1°, which limits their application to only very large catchments. They nevertheless allow for a reasonable simulation of actual flows in such catchments, such as the Danube or other major rivers.

These models can also be used for simulating the impacts of climate change on the flow of most of the main rivers of the planet. Figure 10.6 shows, for example, time series of simulated annual flow anomalies for both the Danube and the Lena rivers and the comparison with the nearest actual measurement station. Observed time series for the actual period unfortunately are generally rather short and/or are not corrected for anthropogenic influence (such as dams and irrigation), which makes it difficult to detect the fingerprint of climate change solely from observations. Nevertheless, simulated time series show a tendency to increasing flow of the Lena, which is compatible with the latest observations and can be explained by an increase in precipitation over its overall catchment area. One can also note a long-term tendency for the Danube flow to decrease. However, this is neither observed nor simulated during the twentieth century, but it could be due to an increase in simulated evapo(transpi)ration. It is nevertheless worth noting that the evolution for the twenty-first century differs notably from

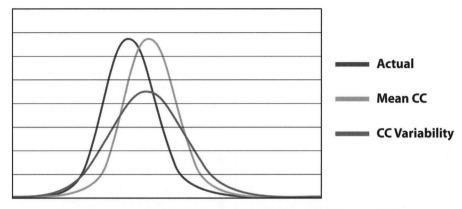

Figure 10.7 Schematic representation of the variation of extreme events probability in the case of climate change (by variation of the surface below the various curves to the right of a given threshold): the 'Mean CC' curve corresponds to a change in the mean value without any change in the variability around the mean, whereas the 'CC Variability' curve accounts for a possible modification of the distribution for the same mean value.

that occurring during the later part of the twentieth century, indicating that a simple extrapolation of observed river flow is not sufficient to assess future runoff.

Impact on extreme hydrological events

Climate change will affect not only the mean state, but it is quite likely that it will also modify the frequency and intensity of extreme events. As sketched in Figure 10.7, it is obvious that a modification of the mean value of a given parameter is de facto leading to a shift in the distribution curve, whose effect is to modify, sometimes very significantly, the probability of occurrence of events corresponding to the distribution tail. Such modifications would of course be enhanced if the distribution curve itself were to change by, for example, an increase in its positive or negative skewness.

As discussed above for Europe, the likely mean increase of winter precipitation and mean decrease of summer precipitation could lead to an increase, both in frequency and magnitude, of apparently contradictory events, such as winter floods and summer droughts. On the other hand, any modification in the precipitation *regime* (such as the respective importance of convective – intense and sporadic – versus stratiform – continuous and moderate – precipitation), and consequently in the frequency and intensity of extreme hydrological events, is much more difficult to predict than the average conditions. This is due, to a very large extent, to the uncertainties that still affect the modeling of small-scale features (both in time and space) of the atmospheric water cycle. One approach to looking for a possible hydrological signal linked to climate change is to simulate future climate with a fairly large increase in atmospheric CO_2 concentration, so that the anthropogenic signal can become larger than the noise due to natural variability noise and/or model uncertainties. Quadrupling the atmospheric CO_2 concentration leads to a very significant increase in large floods (those presently characterized as 'centennial' floods) for high-latitude rivers with large enough (larger than 200,000 km^2) catchments. The same simulations also indicate that this increase in winter flooding might appear as early as the beginning of the twenty-first century. Such results would nevertheless need to be confirmed by other modeling studies.

Atlantic cyclones

Cyclones are very intense climatic phenomena that are associated with very strong winds, intense precipitation, and very high ocean surges. They can be very devastating for populations. For instance, 138,000 people died in 1991 from cyclones in Bangladesh. They are further very damaging for infrastructures: it is estimated that the losses due to cyclone Andrew reached $30 billion, while those due to Katrina reached $80 billion. Cyclones also deeply affect many coastal ecosystems and may lead to long-standing ecological perturbations.

Cyclones are thermodynamic phenomena that are fed by energy from the ocean surface: the warmer the sea, the more evaporation and the more convective clouds. Once the incipient atmospheric depression is formed, the cyclone gains its energy through the heat released by the condensation of water vapor in clouds. It is known from experience that cyclones preferably develop when the sea surface temperature (SST) exceeds a threshold of about 26.5°. The vertical thermal structure of the atmosphere is also of importance, as is vertical wind shear, which plays an inhibiting role on cyclone formation.

About 70–80 cyclones develop each year over the ocean tropical belt, among which 10 or so affect the Atlantic. It must be noted here that 2005 has been a very special year, with no less than 26 cyclones forming and developing over the Atlantic. It is then of major importance to assess whether the number (and possibly the intensity) of cyclones is going to steadily increase as climate change takes place, or on the contrary, if the intense cyclonic season that occurred in 2005 is simply a natural fluctuation in the annual number of cyclones. A controversy developed in the scientific community with respect to the above alternative. As a matter of fact, the 2006 season had many fewer hurricanes than predicted.

Observing the cyclonic activity

Cyclones can be observed either directly, through meteorological measurements, or indirectly, through economic evaluation of their damages.

Direct observation is hampered by the fact that satellites did not exist before the 1970s to identify cyclones from space. Observation from the surface was the only way to count cyclones, and many of them could be missed when they affected unpopulated areas (e.g. when they did not land over an inhabited coast). Reconstruction of long-time series of cyclonic activity necessary to characterize a possible change in the distribution (see the above section on extreme events) is, then, quite difficult and may lead to results biased toward an increase in the number of cyclones. Bias correction is not straightforward, as inferred from the controversy that developed between some well-known specialists.

The intensity of a particular cyclone can be evaluated through a 'power-dissipation index' (PDI), which is built mainly from the observed maximum wind speed. A significant correlation has been observed since the 1970s between the SST and the PDI: an increase of 2° in the SST corresponds to a 40%–50% increase in the PDI. It would then seem that recent observations are consistent with an increase in the intensity of cyclones.

It has, however, been argued that an increase in the intensity of cyclones, as measured from their PDIs, should translate into increased damages. Economic calculations show that damages indeed have increased over recent years, but such calculations have to be corrected for the fact that the damage increase may result from factors unrelated to the cyclones themselves, such as economic inflation, increase in the exposed population, and/or in the value of the exposed properties. When such corrections are included, the trend for increased damages over the recent years disappears. One should note, however, that the way corrections are defined remains a subject of debate, as other factors that would lead to the opposite trend could be taken into account (e.g. preventive measures in the construction of houses, building of dams, better education of the population, etc.).

Direct and indirect observations of the past cyclonic activity consequently cannot lead to unambiguous conclusions about the possible influence of climate change.

Modeling the future

Numerical simulations lead also to quite ambiguous results: some models indicate that the number of cyclones does increase when climate warms, while some others lead to the opposite results. This might be a consequence either of the still-too-coarse resolution of climate models (the size of their numerical grid is large compared with the size of the cyclone itself, so that 'numerical cyclones' are difficult to characterize) or of a poor description of the influence of vertical wind shear, a phenomenon which is known to be difficult to accurately simulate with the present state-of-the-art models. In any case, numerical results are not consistent with a systematic correlation between SST and cyclonic activity.

Climate modeling is presently developed in a number of research groups so as to offer better and more complete models and achieve finer simulations. This is a necessary step for inferring whether or not cyclonic activity is going to be modified by climate change, and if it is, what the response will be, both qualitatively and quantitatively. Knowing the response today is a major challenge in preparing for climate change.

Impacts on vegetation, ecosystems, and agriculture

Natural vegetation and ecosystems, as well as phenological features (e.g. blooming dates for plants and mating dates for wild animals) and productivity of agriculture, are quite well adapted to their present local climate. Furthermore, natural systems and living species are very sensitive to climatic and hydrological changes, so that they can even be used as proxies for such changes.

Past changes in temperature and precipitation regimes have led to modifications in agricultural activities. Dates for wine harvesting, for instance, are excellent proxies for spring and summer temperatures, and this has allowed the qualitative and quantitative reconstruction of past temperatures in Western Europe over many centuries. It is unfortunately impossible with the present state of knowledge to confidently reconstruct other parameters, such as the fluctuations of the hydrological cycle.

Studies of current trends for a number of ecosystems and wildlife indicate that some impacts of climate change can already be noticed: approximately 80% of changes that have been published in the scientific literature for wild flora and fauna are compatible, and even more are coherent with expected effects of climate warming. Phenological phases are taking place earlier (by about a little more than two days per decade) and systems and populations are moving either northward (by about 6 km per decade) or toward higher latitudes (by about 6 m per decade).

Agricultural systems also are sensitive to climate change, but this influence is limited by the fact that man generally acts to minimize the impacts and reach optimal productivity (e.g. by irrigation, antifrost measures, etc.). Thus, although agricultural systems cannot be used as climate proxies, it nevertheless remains that agricultural production is impacted by climate variations. Specialized models have been used to predict the influence of climate change on agricultural production over the coming decades. Although these studies are preliminary, they tend to show that agriculture should have lower productivity if the warming exceeds some threshold. This threshold is small in the tropical regions, where water can be the most important limiting factor. It is slightly larger at mid- and high latitudes, where a moderate increase in temperature might have beneficial consequences.

Such studies cannot be considered to have a real predictive quantitative value because, for example, they account neither for changes in water resources nor for enrichment in atmospheric carbon dioxide, two factors known to greatly influence agricultural productivity. They nevertheless indicate the likely changes to be expected, at least from a qualitative point of view.

Impact on human health

Some of the most pressing questions concerning the possible impact of climate change on human health are whether climate change will lead to changes in the areas affected by vector-borne diseases or whether it will influence the distribution of the so-called emerging diseases.

More than 36 new infectious diseases have been identified by the World Health Organization (WHO) since 1976, many of them reappearing in regions where they had been absent in recent years, as in the case of malaria and dengue. This could be ascribed to a number of factors, including anthropogenic ones, like deforestation, agricultural, industrial and hydrological development, road construction, and airborne transportation, but also to climatic ones, either direct (temperature, humidity, precipitation, radiation) or indirect, through the modification of ecosystems and, more generally, of biodiversity.

Climatic factors are usually considered very effective in triggering new disease outbreaks, as they amplify the influence, transmission, and diffusion of pathogenic agents. For example, in the case of vector-borne diseases, local climatic conditions (rain, temperature, open waters, vegetation, etc.) control the biological properties of the vectors (mosquitoes, ticks, etc.), which, in turn, determine their hosting capacities with respect to the pathogenic agents responsible for the disease. Climate change could lead to an increase in the geographical areas such vectors inhabit and would then favor their development. It is presently estimated that 40%–50% of the world population can potentially be affected by either malaria or dengue. Nevertheless, it remains necessary to understand over which spatial and temporal scales the impacts will be most important.

Impact of mean climatic conditions

Mean climate change will affect the spatial distribution of emerging diseases, through either its influence on biological features of the vectors (e.g. lifetime, reproduction season) or bird migrations, as these animals do host a number of pathogenic agents. It is usually considered that such diseases as West Nile fever and malaria might affect countries like France again, as climate warming may reach 1.0–2.0 °C (1.8–3.6 °F) in winter and more than 2.0 °C (3.6 °F) in summer and fall, with corresponding changes in precipitation (see above). It should, however,

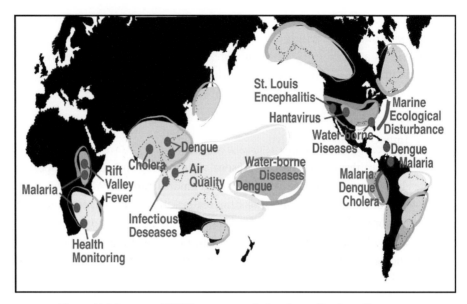

Figure 10.8 Impacts of El Niño events on the breakout of various diseases.

be mentioned here that the rapid multiplication of mosquitoes carrying the virus of the West Nile fever seems to depend more on the frequency of rainy episodes than on the total amount of rain. Climate variability, then, is also an important factor for which to account.

Impact of climate variability

Events exhibiting interannual variability, such as El Niño Southern Oscillation (ENSO), are known to influence significantly emerging diseases, because they correspond with large variations in precipitation amounts from one year to another. Numerous studies have indeed shown the influence on malaria, dengue, cholera, flu, asthma and other respiratory diseases, arboviroses, etc. for many regions of the world like South America, but also for Bangladesh, Australia, and even France. Figure 10.8 indicates the regions where these impacts can be seen during El Niño events.

Impact of extreme events

The frequency and intensity of extreme events such as heat waves (see above) will keep increasing during the twenty-first century. As already noted, the 2003 Western Europe heat wave has killed a few tens of thousands of people, mostly older persons living in large cities, where the urban heat island and pollution effects combine. Quite similarly, a possible increase in the number and/or intensity of cyclones could have both direct and indirect (such as diarrhea) effects on human health.

Some final remarks

The above distinction between three temporal scales of impacts (mean, variability, extreme events) has been introduced for the sake of simplicity, but it cannot be considered fully appropriate because these three scales are interconnected, as local and global factors can have either positive or negative feedbacks.

A second remark concerns the present state of knowledge of the relations between health and climate. Most existing studies can be considered as epidemiologic ones, in which one looks for statistical relationships between diseases and some climatic parameters. It now seems necessary to turn to more integrated and multidisciplinary approaches, based on the search for physical and biological mechanisms that can explain deterministically the breakout of diseases.

Acknowledgments

This author would like to thank many of his colleagues with whom he had very stimulating discussions and who provided very often unpublished papers and notes from which this chapter has been prepared. Thanks are particularly due to Pierre Chevallier, Michel Déqué, Hervé Douville, Jean-Pierre Lacaux, Katia Laval, Serge Planton, and Yves Tourre.

11

Atmospheric chemistry and climate interactions

GUY BRASSEUR AND MARIE-LISE CHANIN

> The earth does not belong to man; man belongs to the earth. All things are connected like the blood, which unites one family.
>
> – Chief Seattle, 1854

> En ce qui concerne l'avenir, votre tâche n'est pas de le prévoir, mais de le rendre possible.
>
> – Antoine de Saint-Exupéry

Introduction

When compared with the atmosphere of other planets in the solar system, the chemical composition of the Earth's atmosphere is unique. The major constituents are not carbon dioxide as on Mars and Venus or hydrogen and helium as on Jupiter and Saturn, but nitrogen and oxygen. The atmospheric abundance of these latter gases is directly related to the presence of living organisms. Nitrogen, for example, is produced as a result of bacterial activity in soils, while oxygen is released as a product of photosynthesis by plants. In addition to these major gases, the Earth's atmosphere contains a myriad of other constituents, whose

> **Guy Brasseur** has been studying the processes that determine the chemical composition of the atmosphere, specifically the impacts of human activities on the ozone layer and on climate. He is an associate director at the National Center for Atmospheric Research (NCAR) in Boulder, Colorado.
>
> **Marie-Lise Chanin** has studied the middle atmospheric structure and dynamics before getting involved in initiating and co-directing the component of WCRP dealing with the role of the stratosphere in climate. Her whole career has been with the Service d'Aéronomie du CNRS, where she is Senior Scientist Emeritus.

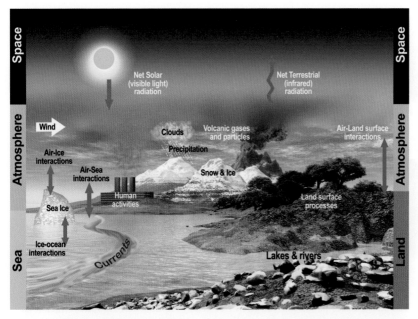

Plate 1 A simplified rendition of the climate system components and their interactions.

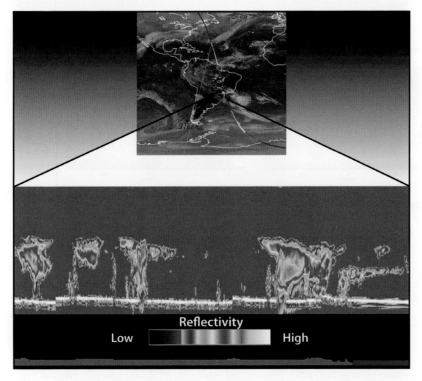

Plate 2 Clouds at CLOUDSAT satellite trajectory overlaid on geographic map (top); cross-section of radar data along CLOUDSAT satellite trajectory over the Andrea tropical storm on May 9, 2007, indicating cloud characteristics expressed in reflectivity (bottom). High values of reflectivity correspond to high water content.

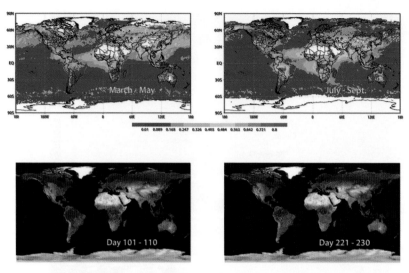

Plate 3 Global distribution of aerosols and vegetation fires for March–May 2005 (left) and for July–September 2005 (right). Regions with open biomass burning appear in red on the lower panels and are characterized by elevated aerosol optical thickness. Other aerosol types, such as dust blowing out of the Sahara, sea-salt aerosols over the oceans and pollution aerosols over industrialized regions, are also visible on the upper panels. All the data are from the MODIS instrument on the NASA Terra satellite. The fire information is from http://rapidfire.sci.gsfc.nasa.gov/firemaps, and the aerosol information from http://lake.nascom.nasa.gov/movas/.

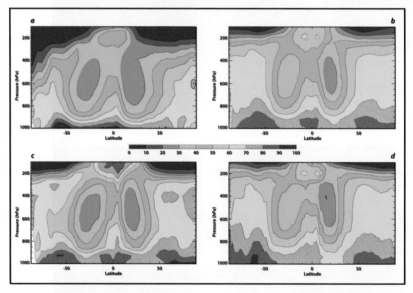

Plate 4 Zonal-average relative humidity (%) as observed by the NASA Atmospheric Infrared Sounder (AIRS) and as computed by various climate models for northern winter months: December, January, and February. (a) AIRS observations over the period 2002–06; (b) average of 19 climate model simulations over the period 1980–99; (c) 1980–99 model simulation by Japan's Frontier Research Center for Global Change; (d) 1980–99 model simulation by the United Kingdom Meteorological Office. (Figure is courtesy of NASA and JPL. The panels were produced by Drs. Thomas Hearty and Duane Waliser of JPL. For more information, see http://airs.jpl.nasa.gov.)

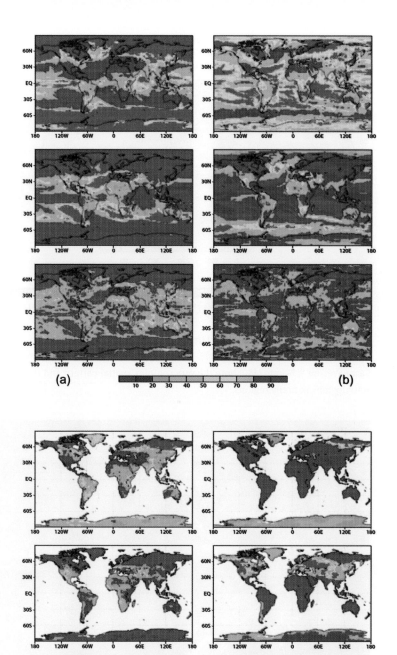

Plate 5 (a) Percentage of IPCC models that predict increased annual precipitation (top panel), decreased precipitation (middle panel), and no change (bottom panel) between PICNTRL and scenario SRES A2. (b) Same as (a), except the variable is annual evapotranspiration. (c) Same as (a), except the variable is annual runoff. (d) Same as (a), except the variable is March–April–May snow water equivalence.

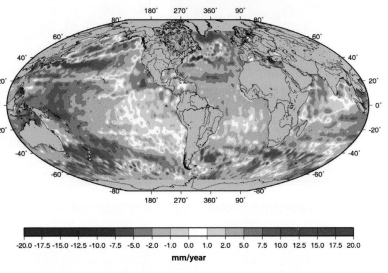

Plate 6 Estimated trend in sea level (mm/y) from altimetric satellite measurements for the time interval from 1993–2005. White regions are areas of no data (after A. Cazenave and S. Nerem [43]). Note the very complex spatial pattern and that sea level actually falls in many regions.

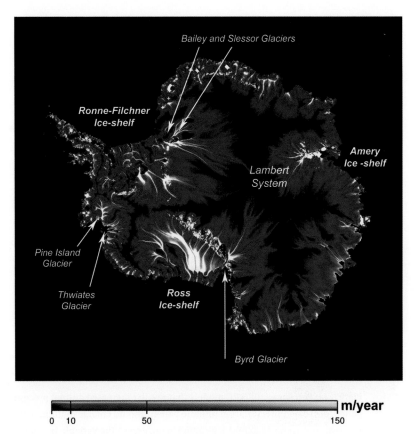

Plate 7 Ice flow in Antarctica: one can observe large ice flows still visible several hundreds of km upstream. Eighty percent of the continental ice are drained by only a few percent of the coast.

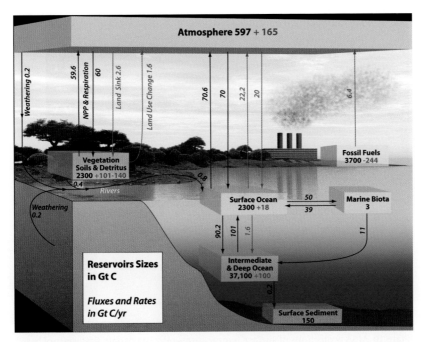

Plate 8 Schematics of the global carbon cycle.

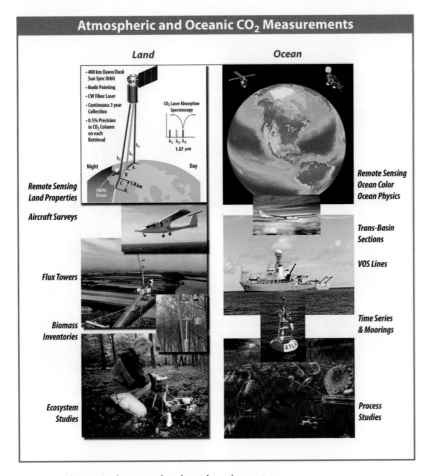

Plate 9 An integrated carbon observing system.

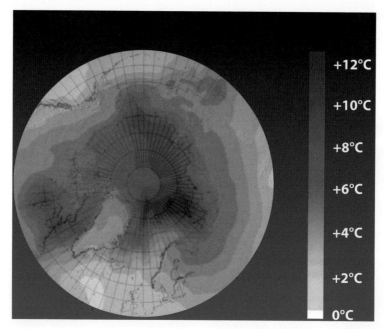

Plate 10 ACIA projections of surface temperatures over the twenty-first century based on the average of the five ACIA computer models and using the B2 scenario (source: ACIA Report [129, 130]).

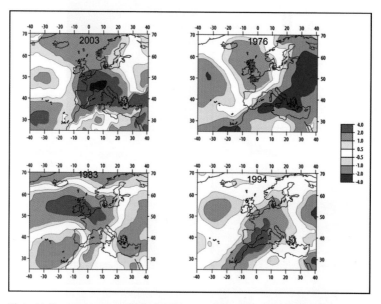

Plate 11 Temperature anomaly (in °C) at 850 hPa during June–July–August 2003 (upper left), June–July–August 1976 (upper right), June–July–August 1983 (lower left), and June–July–August 1994 (lower right).

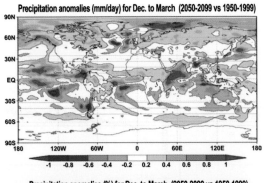
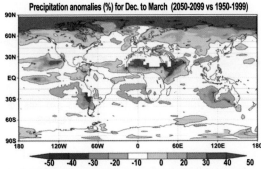

Plate 12 Global precipitation anomalies for the twenty-first century (top: mm/day; bottom: %) as simulated for the December–March period in the case of a B2 scenario with 'Arpège-Climat' model (after M. Déqué).

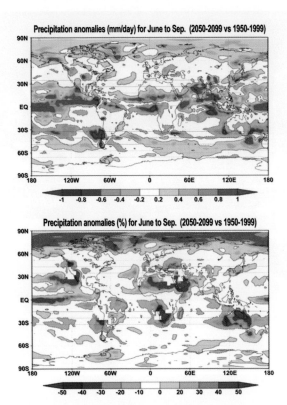

Plate 13 Global precipitation anomalies for the twenty-first century (top: mm/day; bottom: %) as simulated for the June–September period in the case of a B2 scenario with 'Arpège-Climat' model (after M. Déqué).

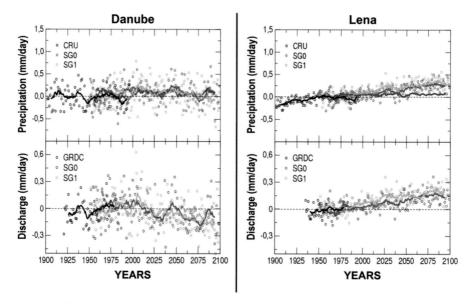

Plate 14 Annual anomalies (circles) and 11-year moving average (curves) for precipitation (top) and discharge (bottom), for both observations (black symbols) and simulation (red symbols), in the case of the Danube (left) and the Lena (right) rivers.

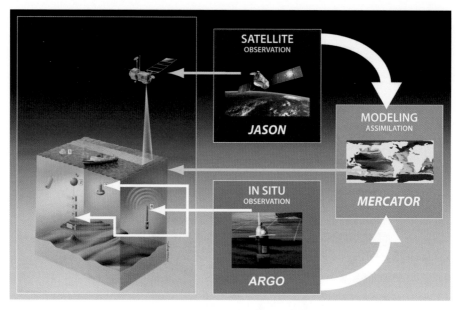

Plate 15 Artistic rendition of an integrated ocean observing system.

relative abundance is generally low, while their potential to affect climate can be very high. This is the case of the greenhouse gases, such as water vapor, carbon dioxide, methane, nitrous oxide, and chlorofluorocarbons. Ozone, a strong atmospheric oxidant that efficiently absorbs ultraviolet light and therefore protects the biosphere (including humans) from harmful solar radiation, is also a greenhouse gas. Other compounds that have no direct influence on climate may play an indirect role. This is the case of the very reactive hydroxyl radical (called OH radical, produced by the oxidation of water vapor in the presence of light), whose atmospheric concentration determines the rate at which some climatically important compounds like methane are chemically destroyed in the atmosphere. The atmospheric concentration of OH is controlled not only by the abundance of water vapor but also by the concentration of other chemical species, including ozone, nitrogen oxides, carbon monoxide, methane, and a large number of non-methane hydrocarbons. A considerable amount of these compounds are produced by human-related activities, and it is important to assess whether the changes in the atmospheric composition play an important role when projecting the evolution of climate in the next centuries.

Many important interactions exist in the climate system (see Figure 1.1), and these interactions can be affected by anthropogenic chemical emissions, land use changes, wildfires, urbanization, etc. Although the action of humans on the climate system is much more complex that the sole emissions of carbon dioxide and other greenhouse gases, one also has to emphasize the role of ecosystems (land and oceans) in the interactions between chemistry and climate. Climate forcing occurs primarily through radiative effects of greenhouse gases and aerosols. In addition to the human-related perturbations that affect the chemical composition of the atmosphere, the atmospheric level of greenhouse gases is determined by several other processes, including the exchanges of carbon and nitrogen with the marine and continental biosphere. Furthermore, the atmospheric concentration of aerosols, which also affects climate through direct and indirect processes (see Chapter 3), is governed by *in situ* oxidation processes involving gas phase compounds, emissions from the ocean surface of sulfur and sea salt, the release of soot from wildfires, mobilization of dust, specifically in the deserts, wet scavenging associated with precipitation, and dry surface deposition processes. In short, no single process related to chemistry-climate interactions can be isolated from the other processes, and a broad approach that recognizes the complex interactions between physical, chemical, and biological Earth system processes must be adopted. Modern climate models must therefore account for this complexity and reproduce all potential feedback mechanisms that could have a long-term effect on future climate.

Atmospheric composition and chemical processes

Over the 4.6 billion years of the Earth's history, there have always been intimate relations between climate and the chemical composition of the atmosphere. The abundance of chemical compounds has evolved in response to geological and biological processes, which themselves have been affected by natural climate changes. Life has played a key role in maintaining the air chemical composition far away from thermodynamic equilibrium conditions. As indicated in Chapter 8, ice records have shown a strong co-evolution of climate cycles, with cycles detected in the atmospheric abundance of long-lived gases such as carbon dioxide and methane. These cycles highlight the strong couplings that exist through the carbon cycle between the physical climate, ocean dynamics, and terrestrial ecosystems.

Since the eighteenth century, the natural evolution of the atmospheric composition has been progressively perturbed by human activities, and for several chemical compounds, the anthropogenic source has become larger than the natural source. A major consequence of the human enterprise has been to increase dramatically the atmospheric concentration of several greenhouse gases. The fate of many of these gases in the atmosphere, and hence their overall atmospheric lifetime, depends on chemical or photochemical mechanisms that take place either in the troposphere – the atmospheric layer extending from the surface to approximately 10 km (30 000 ft) altitude and subject to rapid vertical exchanges and other dynamical perturbations – or in the stratosphere – the dynamically stable atmospheric layer extending from approximately 10 km to about 50 km (150 000 feet) altitude. The major chemical processes that affect radiatively important and chemically active atmospheric compounds are briefly reviewed in the following sections.

Stratospheric chemistry

Even though it includes only 10% of the atmospheric air, the stratosphere contains the ozone layer and therefore plays an important role in the overall heat budget of the atmosphere. Unlike many other gases, ozone is not emitted at the surface but is produced in the stratosphere by the action of solar ultraviolet radiation on molecular oxygen. Once formed, ozone can be destroyed by different fast-reacting radicals, whose atmospheric concentration can be affected by human activities. The major loss is provided by nitrogen oxides, whose major stratospheric source is provided by the oxidation of nitrous oxide. Reactions involving chlorine radicals were invoked in the early 1970s to highlight the fact that chlorofluorocarbons would be a threat to the ozone layer. Early models, however, showed that the related ozone depletion should be limited to a few percent and should take place in the upper stratosphere. The observations in the early 1980s by British and Japanese scientists of exceptionally low ozone concentrations in the Antarctic from September to November suggested that, in the vicinity of the South pole, springtime ozone was being rapidly destroyed in the lower stratosphere, i.e. in the layers where it is most abundant. This dramatic perturbation, known as the 'Antarctic ozone hole,' was later explained by physical and chemical mechanisms that had been ignored so far, and specifically by the fact that the relatively inert chlorine and bromine reservoir compounds were activated and converted to reactive radicals at the surface of tiny liquid or solid atmospheric particles. Thin polar stratospheric clouds form during winter between 15 and 26 km altitude in the particularly cold polar regions. Reactive chlorine, for example, under the form of chlorine atoms (Cl) and chlorine monoxide (ClO), destroys ozone very efficiently. It was also stressed that, due to the long residence time of chlorofluorocarbons in the atmosphere (50–100 years), the level of chlorine would remain high for several decades, and hence the ozone hole would persist most likely until the middle of the twenty-first century.

Bromine-containing halocarbons such as methyl bromide (used in support of some agricultural activities like fruit production) or the halons (used, e.g. in fire extinguishers) have even higher ozone depletion potentials than do many chlorine-containing halocarbons. Laboratory studies show that a bromine atom is approximately 60 times more effective than a chlorine atom in destroying an ozone molecule.

For a long time, the scientific community treated the problem of ozone depletion separately from the climate issue. It has now become evident that the two issues are linked, which highlights why stratospheric ozone needs to be considered in studies that deal with climate change.

Taking into account stratospheric ozone in climate models will also be necessary to account for the effects of solar variability. The variation of the solar radiative energy is particularly large in the ultraviolet part of the solar spectrum. The transmission of ultraviolet light to the lower atmosphere is determined by the abundance of stratospheric ozone, which itself is modulated by solar activity. Recent work suggests that the changes in stratospheric composition and dynamics resulting from ultraviolet solar variability have indirect effects on the troposphere through dynamical and radiative coupling that are not fully understood.

The stratosphere is also occasionally disturbed when considerable amounts of sulfate aerosol particles are injected by explosive volcanic eruptions. This has been the case after the major eruptions of El Chichón in 1982 and Pinatubo in 1991. Sulfuric aerosols scatter a fraction of the incoming solar radiation back to space, which tends to cool the Earth's surface. A cooling of typically 0.5 K was observed during the couple of years that followed the eruption of Pinatubo. These changes are not insignificant when compared to the current warming caused by greenhouse gases. However, volcanic perturbations occur episodically and their effects are limited to a few years. As mentioned in Chapter 3, the injection of sulfuric aerosols in the stratosphere has been proposed as a geoengineering method to oppose global warming. Even though such attempts may prove to be effective in reducing the mean temperature of the Earth, potential side effects, which could be significant and damaging, remain to be quantified.

Tropospheric chemistry

The atmosphere is an oxidizing medium that destroys most of the primary compounds that are emitted by natural or anthropogenic processes (including some greenhouse gases such as methane). Atmospheric oxidation does not occur through direct reactions with the oxygen molecule but by more reactive chemical species, specifically the hydroxyl radical and the ozone molecule. As stated above, the hydroxyl radical (OH) is produced by the oxidation of water vapor by oxygen atoms. For this process to occur, the atom of oxygen must be in an excited electronic state, and such an atom can be formed only following the destruction of an ozone molecule by solar ultraviolet light at wavelengths less than approximately 320 nm. The hydroxyl radical is then converted into a peroxy radical (HO_2) by reactions between OH and carbon monoxide (CO), methane (CH_4), and other hydrocarbons. This radical is then converted back to OH through a reaction with nitric oxide (NO). Thus, the abundances of OH and HO_2 are strongly affected by the atmospheric concentration of gases like carbon monoxide and nitric oxide, whose atmospheric abundance is often strongly enhanced as a result of anthropogenic emissions. A product of the reaction between HO_2 and NO is nitrogen dioxide (NO_2), which can be dissociated by solar radiation. This photochemical process leads to the formation of ozone and therefore plays a key role in the troposphere. Under most circumstances, the production rate of tropospheric ozone is controlled by the atmospheric level of nitric oxide, and because NO is, to a large extent, a product of fossil fuel combustion and biomass burning, the level of surface ozone is directly affected by human activities.

It is believed that the surface concentration of ozone in industrialized areas has increased by at least a factor of two since the preindustrial era. By how much has the level of OH, and hence the oxidizing capacity of the atmosphere, changed since the preindustrial era? This question remains unresolved. Photochemical smog events with high ozone levels and potentially severe health effects are observed in the vicinity of intense sources of nitric oxide, carbon monoxide, and hydrocarbons (called ozone precursors) during meteorologically stable situations. Will the frequency of such pollution events change in response to climate change in the future? The answer to this question is also unknown.

The formation of atmospheric aerosol particles results in large part from *in situ* chemical and microphysical transformations that occur in the troposphere. Sulfate aerosols, for example, are produced by the conversion of sulfur dioxide (SO_2), a product of coal combustion, by hydrogen peroxide (H_2O_2) and ozone (O_3) molecules dissolved in water droplets. Organic aerosols, which are particularly abundant in the tropics, result in large part from the oxidation of volatile organic compounds, such as the terpenes emitted by vegetation.

A region of the atmosphere in which chemical compounds exert a particularly important influence on the radiative forcing of the climate system is the tropopause, the thin atmospheric region that separates the troposphere from the stratosphere near 8–16 km altitude (25 000–50 000 feet). This layer can be regarded as a dynamical barrier, protecting the stratosphere from excessive upward transport of tropospheric water and limiting the downward intrusion of stratospheric ozone into the troposphere. The properties of this layer, in which interactions between radiative, dynamical, chemical, and microphysical processes are complex, could be modified as a result of climate change. Much needs to be learned about climate forcing by observing the transport and transformations of chemical species, the influence of convective motions and lightning particularly in the tropics, the formation and fate of cirrus clouds, the role of mesoscale disturbances at mid-latitudes, the development of monsoon systems, and other processes in this thin layer of the atmosphere, which is not well monitored from space or easily accessible to airborne observing platforms.

The role of human activity

Chemical emissions

The industrial development of the last century combined with population growth has contributed to substantially modify the chemical composition of the atmosphere. As humans started to produce large quantities of chemical compounds, they did not always realize that the related emissions would modify the conditions that had been sustained over the past millions of years.

Methane (CH_4) is released to the atmosphere as a result of microbial activities in anaerobic (oxygen-deficient) environments, specifically in wetlands, lakes, marshes, and rice paddies. It is also produced by the digestive system of ruminants. Leaks in the production facilities and transport systems of natural gas as well as coal mining activities have added to the emissions of biogenic methane. Important quantities of methane are also trapped in the Arctic under the permafrost and could be released into the atmosphere as climate warms with, as a consequence, a substantial enhancement of the greenhouse effect. Methane is slowly destroyed by the hydroxyl radical and to a lesser extent by chlorine atoms. This destruction process gives rise to two molecules of water vapor and therefore enhances the concentration of another greenhouse gas, particularly in the middle and upper stratosphere.

Recent studies suggest that water vapor in the low stratosphere (around 20 km altitude) has been significantly increasing in the last 35 years, while other observations, mostly from satellite sensors, suggest more complex variations as a function of time. More systematic measurements are needed to provide more reliable information about possible water vapor trends in the middle atmosphere. The long-term evolution of methane can be observed in ice cores extracted from Antarctica or Greenland (see Chapter 8). During the last hundreds of thousands of years, and until the eighteenth century, the concentration of methane has fluctuated between approximately 350 and 700 ppbv in phase with temperature (climate) changes. After that period, the atmospheric abundance of methane started to increase at the same time as population growth became dramatic and agricultural practices more intense. The present atmospheric concentration of this gas (1700 ppbv) is more than twice what it was in 1750. Despite the fact that little growth has been observed in the last two decades, the atmospheric concentration has reached a value never observed in the last 700 000 years.

Nitrous oxide is released to the atmosphere as a product of nitrification and denitrification processes triggered by bacteria in soils. It is the major source of nitrogen oxides in the stratosphere, which in turn determines to a large extent the rate at which ozone is destroyed between 15 and 35 km altitude. The concentration of nitrous oxide has been increasing in the troposphere at the rate of 2%–3% per decade in the last 25 years, probably as a result of the more intensive use of nitrogen fertilizers for agricultural processes.

Nitrogen oxides (NO and NO_2) are produced primarily as a result of combustion processes, although some natural sources (emissions by soils and wildfires and production by lightning strikes in thunderstorms) are also important. Their lifetime is very short (several hours or days) in the lower atmosphere, so that they do not accumulate in large quantities, but they are very reactive. As stated earlier, these compounds play a major role in the formation of tropospheric ozone and in the destruction of stratospheric ozone. Nitrogen oxides are also released by aircraft engines, primarily in the upper troposphere and lower stratosphere. The effect of aircraft emissions of a projected fleet of supersonic aircraft led to intense controversies in the early 1970s. The question was also raised in relation to the operation of the Concorde that had started a few years earlier. In the 1990s, much attention was given to the effects on ozone and climate of the existing fleet of commercial subsonic aircraft. Even though the environmental effects of aircraft operations were estimated to be relatively limited (but not negligible), this issue motivated the aeronautical industry to develop cleaner engines. An emerging issue is the emissions of nitrogen and sulfur oxides by the engines of ships. These are not strictly regulated, and they release considerable amounts of chemical pollutants in the relatively clean oceanic regions. The growing intervention of mankind in the rate at which molecular nitrogen N_2, the most abundant atmospheric gas, is converted into other forms of nitrogen has pushed the nitrogen cycle far away from its equilibrium state in all compartments of the Earth system. Oxides of nitrogen, when deposited on soils, are powerful fertilizers of the biosphere and consequently affect the rate at which carbon is captured from, or released into, the atmosphere. Thus, here again, the manipulation of the nitrogen cycle by humans to enhance food production has an indirect effect on the carbon cycle, and therefore on climate forcing.

The atmospheric concentration of ozone (O_3) has been measured at the Earth's surface for more than a century. The earliest systematic and continuous observations were made at the Observatoire du Parc Montsouris in the south of Paris in the late 1800s and early 1900s. Even though the measurements made at that time were not very accurate and probably influenced by several other factors including the relative humidity, the values reported were considerably lower than those determined at present by more elaborate techniques. As stated above, the present concentration of surface ozone in industrialized areas is at least a factor of two larger than during the preindustrial era. The rate of increase is highly variable in space, depending on the level of urbanization and industrialization, and is therefore very different in both hemispheres.

The environmental impacts of halocarbons (including the chlorofluorocarbons, or CFCs) are also important. Most of these compounds that are presently detected in the atmosphere did not exist in the preindustrial atmosphere but started being produced in considerable amounts by industry in the second half of the twentieth century. They were developed for a variety of applications, including their use as refrigeration agents in air-conditioning systems, propellants in aerosol cans, cleaning agents, etc., because of their chemical stability and their safety for users. The atmospheric lifetime of chlorofluorocarbons can reach 50–100 years. The only significant destruction mechanism is the slow photodestruction of these molecules by solar ultraviolet radiation when they reach the stratosphere after several years. This destruction process liberates chlorine and in some cases bromine atoms, which have the potential to destroy ozone molecules through catalytic cycles (i.e. each single chlorine or bromine atom has the potential to destroy thousands of ozone molecules). Halocarbons are radiatively active and therefore contribute to the greenhouse effect, as highlighted in Chapter 2. Since they were identified as the cause of ozone destruction, the production of several halocarbons has been discontinued as a result of the Montreal Protocol signed in 1987 and of further strengthening amendments adopted in the early 1990s. Alternative halocarbons, with considerably less ozone destruction potential, are currently produced to replace the traditional CFCs. These new compounds may have significant greenhouse warming potentials as well as rather long atmospheric lifetimes. Their production and use are now under control so that their climatic impact will remain limited [134].

Bromine and chlorine compounds are produced not only by industrial processes. Natural sources (e.g. oceanic sources of methyl bromine and methyl chlorine, volcanic sources of hydrogen chlorine) are significant and represent 16% and around 40%, respectively, of the anthropogenic production of chlorine and bromine.

In the stratosphere, the ozone layer has been subject to careful monitoring, mostly since the discovery of the ozone hole in Antarctica. Since the early 1980s, the total ozone amount has been decreasing by about 3% and 6% at mid-latitudes in the North and South Hemispheres, respectively. In the Antarctic and Arctic, the springtime ozone decrease over the same period has been as large as 45% and 25%, respectively. Detailed studies have demonstrated that these changes resulted primarily from the release in the atmosphere of the chlorofluorocarbons. With the phasing out of these products as a result of the Montreal Protocol, the concentration of ozone-depleting substances has been gradually decreasing in recent years. However, the rate at which chlorine is eliminated from the stratosphere is slow; it is determined by the long chemical lifetime of these CFCs.

In spite of the measures taken to reduce the chlorine burden of the stratosphere, it is necessary to exert a tight control on the use of these products. Large stocks of halocarbons still exist and are being used illegally in certain parts of the world. International cooperation is required to ensure that the products listed in the Montreal Protocol and its subsequent amendments, as well as in the UNFCCC, are effectively and entirely banned from usage. It is interesting to note that the radiative forcing provided by the halocarbons represents 13% of the overall forcing from the well-mixed greenhouse gases.

Since 2000, a pause has been observed in the ozone decrease, which may be interpreted as the first sign leading to the ozone recovery that the models had predicted. It would be particularly interesting to notice some initial ozone recovery in the polar regions. Unfortunately, the large dynamical variability in the Arctic makes it difficult to document such a change. In the Antarctic, where the ozone hole is less variable, a return to a full recovery, i.e. the situation observed before the 1980s (if it ever does happen), should not manifest itself before 2040–2060. It is not known if climate change, and specifically changes in stratospheric temperature and circulation, will accelerate or delay the recovery of ozone, especially in Antarctica.

As stressed in Chapter 2, there is growing evidence that the increase in atmospheric concentration of greenhouse gases has been the major source of the warming of the troposphere and of the Earth's surface. The long-term increase in the temperature is due mainly to the radiative forcing provided by CO_2 and methane, but the additional effects of the other greenhouse gases and the direct and indirect effects of aerosols cannot be ignored. In addition, the observed decrease of ozone in the lower stratosphere has produced a cooling effect at the surface, but this contribution has been estimated to be rather small, except in the center in Antarctica, where a significant cooling is attributed to the reduction in stratospheric ozone (see also Chapter 7).

The changes in the composition of the atmosphere that are causing a warming of the lower atmosphere are producing a simultaneous cooling of the stratosphere. The explanation is simple: in the upper atmosphere, where the density of gases is considerably lower than in lower layers and the atmosphere is cloud-free, increasing concentrations of greenhouse gases with time tend to enhance the radiative infrared emissions to space; at the same time, depleted concentrations of ozone reduce the amount of solar ultraviolet energy that is absorbed in the stratosphere. Models have derived a cooling that is quite in agreement with the analyses of the last 25 years of data; a cooling of 0.5 K per decade at 20 km reaching 2 K per decade at the stratopause around 45 km altitude [135]. Higher in the atmosphere, in the mesosphere (60–70 km), observations (measurements by rocket-borne instruments and ground-based lidars, optical detection of airglow emissions) show a cooling that could be much larger: up to 5–10 K per decade at mid- and high latitudes. This remains to be explained, because models do not reproduce such a large temperature trend.

Impact of climate changes on the chemical composition of the atmosphere

The preceding sections have highlighted the direct and indirect effects of chemical compounds on climate. However, the interactions between climate and atmospheric chemistry are more complex, because the chemical composition of the atmosphere itself is influenced by climate change (see, e.g. [136]). Thus, a number of amplifying (positive feedbacks) or attenuating (negative feedbacks) mechanisms could result from these two-way interactions. Unfortunately, the quantitative importance of the chemical feedbacks in the climate system is not yet established, and it is difficult to conclude that their importance will be substantial in future climate conditions. The concern is that unknown or poorly investigated processes could lead to other processes that would amplify small, human-driven perturbations to unexpectedly large and potentially dangerous effects.

Several processes could be involved. First, in a warmer and wetter climate, the level of OH is expected to increase, with a potential reduction of the atmospheric abundance of ozone precursors such as carbon monoxide and nitrogen oxides. Thus, the effect of climate change would be to reduce tropospheric ozone levels. However, other processes could act in an opposite way, because under warmer conditions the emissions of some ozone precursors (i.e. volatile organic compounds by plants and of nitric oxide by bacteria in soils) could be significantly higher. In addition, if the frequency and intensity of lightning strikes were enhanced in a warmer climate, larger quantities of nitric oxide would also be produced in the middle and upper troposphere, especially in the tropics. As a result, tropospheric concentration of ozone should be larger, particularly in the tropical regions.

Changes in weather patterns resulting from climate change could directly affect air quality. Chemical pollutants are known to be very sensitive to meteorological parameters such as winds, temperature, humidity, boundary layer ventilation, and precipitation. The anomalously hot and stagnant air masses that prevailed during the summer of 1998 in the Northeastern United States and the 2003 heat wave in Europe were associated with exceptionally high ozone concentrations in these regions. In general, daytime ozone concentrations correlate well with temperature, suggesting that surface air quality will be affected by climate change during the next decades.

A number of possible interactions between the chemical composition of the atmosphere and biogeochemical cycles have not yet received sufficient attention. For example, the deposition of chemical substances, which may be altered in a changing climate, affects the input of nutrients to the marine and continental biosphere, and hence the exchanges of carbon with the atmosphere. The input of nitrogen to soils and of iron to ocean by dust particles provides two examples of such interactions. Plant growth should also be affected by ozone deposition. The extent to which the atmospheric abundance of CO_2 and hence climate forcing are affected by such processes remains to be established.

The continuing cooling of the stratosphere expected from increasing atmospheric CO_2 abundance in the future will have some influence on the level of stratospheric ozone. First, it is well-known that the chemical production and destruction rates of stratospheric ozone are quite sensitive to temperature. This may lead to a slowdown in the ozone destruction at mid-latitude. Yet the reverse is expected to occur in the polar regions, where the chemical activation of chlorine occurs on the surface of stratospheric cloud particles, which are formed when the temperature falls below a certain threshold. Second, the cooling of the stratosphere in the decades ahead will lead perhaps to a strengthening of the meridional circulation of the middle and upper atmospheres. This process could accelerate the recovery of ozone at high latitudes, since chlorine-poor and hence ozone-rich air masses originating from low and mid-latitudes would more easily fill up the polar regions, where ozone is rapidly destroyed by chlorine and bromine during late winter and early springtime. Which will be the dominant of these processes, and how will this coupling process between stratospheric ozone and temperature affect future climate? The problem remains open, but is discussed with considerably more details in the latest WMO-UNEP Report on Ozone Depletion [137].

Perturbations in the stratospheric temperature and circulation may affect the circulation of the troposphere and thereby weather and climate. The study of the connections between the troposphere and the stratosphere is a research area in full expansion, and the dynamical connections between the two atmospheric regions have recently attracted a lot of attention. The extent by which stratospheric signals provide the potential for early detection of changes in weather may open new and exciting potentials for improved meteorological forecasts and perhaps for seasonal climate predictions. For example, there are indications that the state of stratospheric dynamics in winter may provide predictive information about the positioning of anticyclonic features over Europe in spring. Thus, it seems important for future meteorological forecast models to include a detailed representation of the stratosphere.

Lessons learned

It is now recognized that the response of the Earth system to human activities is complex and involves multiple interactions between physical, chemical, and biological processes as well as mass, momentum, and energy transfer between the atmosphere, ocean, and land surface. Two-way interactions between the chemical and the climate systems must therefore be taken into account in models that attempt to describe the effects of human activities on the fate of the planet. Changes in the chemical composition do not only produce perturbations in the climate system; they also lead to a degradation of air quality, with impacts on human health and damages to ecosystems. This explains why air quality has become an increasingly important issue for societies and why its control has taken such a high profile in environmental regulations. Stratospheric ozone depletion has provided a prominent example of a global environmental problem that required immediate action from policy makers. This problem has been adequately addressed by effective scientific research in support of a political process that led to the adoption, by several nations, of the Montreal Protocol. As a result, the production of the most dangerous ozone depletion substances was interrupted. The process that led to the formulation and the adoption of this protocol should be assessed in detail because it could serve – at least in part – as a model to address other environmental issues in the future.

The parallel with the protection of climate is obviously far from being straightforward. The scope of the problem, the number of actors involved, and the economic consequences of the potential measures to be taken are considerably larger than in the case of the ozone question. In addition, the control of a reduction in greenhouse gas emission is much more difficult to achieve than in the case of ozone-depleting substances. The concept of an international agreement that is regularly amended when new scientific information becomes available, as in the case of the Montreal Protocol, should be retained for other environmental issues, including climate change.

Another lesson should be taken from the example of stratospheric ozone. In the first half of the 1980s, even though the research community was actively involved in these questions, the discovery of the Antarctic ozone hole came as a complete surprise. Clearly, the observation of the rapid and unpredicted ozone depletion near the South Pole could not be explained without introducing new chemical concepts. Nature has not finished surprising us, even though scientists should be proud of their findings and their ability to address complex questions that are so important for the fate of humanity.

12

Observing system for climate

HELEN M. WOOD AND JEAN-LOUIS FELLOUS

Climate is what you expect, weather is what you get.
– Robert A. Heinlein, *Time Enough For Love*

Un climat pour lui seul : ses plus proches voisins
Ne s'en sentoient non plus que les Américains.
– Jean de La Fontaine, *Jupiter et le Métayer, Fables*

Introduction

Five decades ago, climate research was aimed at describing the mean state of our atmosphere and ocean. Minimal attention was paid by climate physicists to the role of the terrestrial biosphere, and most of the work consisted of laboriously establishing 30-year averages for a variety of atmospheric and other variables, measured with unsophisticated instrumentation. Climate studies were applied to weather prediction, civil engineering, agriculture, water resource management, warning systems, or air and ocean navigation.

Helen M. Wood has extensive experience in national and international efforts to improve collaboration in Earth-observing systems planning and operations. She Manages GEOSS activities for the National Oceanic and Atmospheric Administration (NOAA) and is co-chair of the US Group on Earth Observations. She served as Secretariat Director for the ad hoc intergovernmental Group on Earth Observations from its inception in 2003 until September 2005. Previously, she directed NOAA satellite data processing and distribution operations. She is a fellow of the Institute of Electrical and Electronics Engineers.

The views expressed in this chapter are her own (and those of her co-author) and do not necessarily represent those of NOAA or the US government.

Technology was slow to change over the first half of the twentieth century, which led to slow improvements in accuracy. The whole point of minimizing time-dependent observing biases due to calibration, spatial and temporal sampling, etc., was not captured early on. Moreover, little effort was devoted to improving the accuracy of the measurements, since the general belief was that of an overall stable climate. According to paleoclimatologists, the next glacial advance was still expected to be millennia away.

Why do we need a climate observing system?

Sitting somewhere between applied research and an academic exercise of fluid dynamics, climatology became a major research field at the same time our perception grew that human influence on the planetary environment had reached an unacceptable level. The raised consciousness that mankind has altered and is altering the Earth's climate has changed climatology. Demand grows for a deeper understanding of climate drivers and outcomes. At what rate is climate changing? Is climate change accelerating? What are the respective extents of human-induced and 'natural' causes? What will the climate be like 50 years from now? What are the long-term effects of human influence on the planetary environment? How can we best prepare for and even adapt to changes in local and regional climates? These questions do not come only from the science community, but they are increasingly posed by policy makers and society at large. How these questions are addressed can have major economic consequences for both the developed and developing world.

A global observing system for climate would allow us to answer such questions by fundamentally improving our understanding of the Earth's climate system, including the ability to predict climate change. This knowledge would support decision making to both mitigate and adapt to climate change and variability. The ultimate promise of such a system is to inform sustainable economic and societal development, with minimum perturbations on the climate system.

In climate, as in every science, measurements and observations are required to validate (or contradict) assumptions and theories. But the Earth's climate is not controlled as in a laboratory experiment. Natural variability and turbulence characterize the behavior of the fluids that envelop the Earth, making it difficult to forecast their long-term evolution. Consider the trajectories of two identical balloons released simultaneously from close places, or from the same place at two quasi-instantaneous times: they initially are more or less parallel, then slowly diverge, and after a few minutes, their motions seem totally uncorrelated. The same is true for buoys floating on the sea surface, although at different paces and over different timescales.

Meteorologists use numerical models simulating three-dimensional processes to forecast the weather. These models are based on an a priori knowledge of the physics of fluids synthetically represented by classical equations, such as conservation of mass, energy, and momentum. Given some initial and boundary conditions (including hydrology and land cover) for the system under study, these equations in principle describe its evolution. However, the forecast tends to drift away from the actual weather, due to tiny errors in initial conditions and computational approximations, and after a few days becomes totally unrealistic. This forecast 'drift' makes it necessary to constantly acquire large amounts of new data at all spatial and temporal scales. Today, these data are continuously assimilated into models. This helps to minimize the differences between the predictions of the models and what actually happens in the real world.

In fact, the requirements for an adequate climate observing system are far more stringent than those being used by the meteorological community. Climate analysis and prediction pose different problems, because measuring the small changes associated with the postulated climate change is far from being a simple task, especially as the time horizon for prediction is from decades to a century and more.

Although statistical evidence is recognized by the Intergovernmental Panel on Climate Change (IPCC), getting definitive answers from data collected in the past has been problematic, because recorded temperature time series only began in 1860, and the global temperature average and its variation derived from this disparate data set has been disputed. Questions have been raised concerning the biases attributable to local effects, such as urban heat islands or station displacement, or to changes in measurement techniques of sea surface temperature and to recent inclusion of satellite data. As a matter of fact, all of these uncertainties make it difficult to better understand how the climate system works. Yet building new measurement systems using modern technology also presents a daunting challenge when it comes to detecting surface temperature trends of .01° over a decade or variations in the solar constant as small as one thousandth over the same period of time.

The quality challenge

Natural climate variability occurs on timescales ranging from years to hundreds of millennia. The human-caused (anthropogenic) climate perturbation is superimposed on this natural variability or may simply modify it in ways that are still to be understood. Climate change signals can be detected only if they are greater than the background natural variability.

Assessment of small climate changes through observations over long time periods implies the availability of time series that are both accurate and stable. Here, accuracy is measured by the systematic error of the data with respect to the truth, once random errors have been averaged out. Stability refers to the long-term evolution of the measurement accuracy, perhaps over a decade. Climate change is expected to be 'small' over global scales, while regional changes may be quite large, such as retreating glacier edges, desertification, or coastal zones submersion. In fact, a 1-m rise in sea level is not quite so small, nor is a 5 °C (9 °F) increase in the global mean surface temperature over 100 years (the difference between today's temperature and that of about the 18 000-year age, when the Earth was in full glaciation). Nevertheless, detecting global climate change is much more demanding than monitoring regional impacts. Moreover, detection alone is not enough: observations should be accurate and stable enough to prove beyond any doubt that climate change is occurring and to allow the forcings and feedbacks involved to be fully evaluated.

This has many implications for the characteristics of an observing system able to deliver climate-quality data. Measurements should be taken with accurate, calibrated instruments, converted into geophysical data, quality-controlled, and stored in standard format. Data sets should be precise enough for the early detection of trends over the next decade; homogeneous in location, time, and method; uninterrupted and long enough to resolve decadal trends; and have sufficient coverage and resolution to permit a description of spatial and temporal patterns of change. Needless to say, this is not an easy set of requirements, and the answer to the question of the full adequacy of current climate observing systems to facing this challenge is clearly negative.

The two faces of technology

It is well-known that technology, like the Roman god Janus, has two faces. On the one hand, it has been used by humankind as a tool to exploit our natural environment to a dramatic extent. Yet at the same time, technology has also provided us with tools to observe and monitor the Earth system. Without it, we could not detect on a yearly basis the tiny changes in atmospheric composition or ocean currents or map global land cover changes that factor into our changing climate. Most significant advances in technology for Earth sciences have affected the observing systems, notably the advent of satellites, but also a variety of new sensors able to provide accurate measurements in a hostile environment, such as long-lived profiling floats in the ocean. Advances in computing power, in turn, have allowed the development of more realistic models of the Earth system and its atmospheric, oceanic, and terrestrial components. In addition, mathematical techniques have allowed the timely use of data to improve the quality of forecasts.

Among others, progress in electronics miniaturization, antenna design, communication rates, microwave radar techniques, and stability of oscillators have made it possible to fly active sensors on board remote sensing satellites capable of acquiring all-weather, day, and night observations and to transmit to the ground huge amounts of data with ever-increasing resolution in space and time. Progress in detector sensitivity has allowed atmospheric sounding of moisture and temperature profiles and the detection and monitoring of minor atmospheric constituents.

Progress in satellite systems

Since the first 'beep beep' emitted by Sputnik-1 in 1957 that inaugurated the International Geophysical Year, a lot of progress has been made in satellite systems designed for Earth observation. TIROS-1 (Television Infrared Observation Satellite), 'essentially a black-and-white television set with a camera attached,' launched on April 1, 1960, was the first weather satellite, sending back to Earth daylight pictures of clouds. This US satellite series improved progressively with the addition of an infrared sensor, capable of roughly measuring surface temperatures, and in the 1970s it carried new sensors able to make measurements in five different wavelengths. In 1979, Europe launched its first geostationary meteorological satellite, Meteosat. The following two decades saw the development of an operational suite of highly optimized weather satellites, owned by the United States, Europe, Russia, Japan, China, and India, comprising two series of several platforms each, respectively circling the planet in geostationary and pole-to-pole ('polar') orbits.

In parallel with this effort devoted to meteorological observations, space agencies initiated a number of experimental missions, with a view to extend the capabilities offered by remote sensing to ocean and terrestrial observation, as well as to atmospheric chemistry. For example, the European Space Agency's ERS-1 and -2, launched in 1991 and 1995, were pioneering satellites for Earth observation. Over the last twenty years, a wealth of new techniques has flourished: some of them the extrapolation of observations from ground-based to space-borne sensors, some derived from innovative sensors initially conceived for the exploration of Venus or Mars, and others representing genuine new concepts, fully exploiting the characteristics of in-orbit observation, with no equivalent from the ground or from aircrafts or balloons.

A major breakthrough was made with the advent of active sensors, capable of illuminating their targets with a radar pulse, while most passive sensors were essentially dependent on sunlight or had to be sensitive enough to detect the faint infrared or microwave emissions from the atmosphere and surfaces. Radar altimetry has proved to be a powerful technique that has revolutionized oceanography with its ability to accurately monitor ocean currents and sea level fluctuations.

An incomplete list of the climate variables accessible to space-based passive measurements (and sometimes only to those measurements) includes solar irradiance (total and spectral), Earth radiation budget components (net incoming solar radiation, outgoing longwave radiation, cloudiness), atmospheric temperature, water vapor, ozone, aerosols, precipitation, carbon dioxide, methane, vegetation and forest cover, snow cover, sea ice, sea surface temperature, ocean color (related to phytoplankton surface concentration), lake areas, glacier outlines, etc. Active sensors give access to ocean currents, sea and lake levels, ocean surface wind vectors, sea state, soil moisture, cloud top heights, precipitation rates, ice sheet topography, etc.

Of course, satellite observations have limitations: electromagnetic waves do not penetrate far below the ocean or ground surface; also, some lower atmosphere parameters are practically out of reach from space observation due to atmospheric opacity; satellite lifetimes are rarely longer than a few years; instruments degrade in the harsh orbital environment; and calibration of sensors is often difficult and unstable. No space observation is self-sufficient, and satellite data must be complemented and validated by *in situ* measurements.

It is worth noting that, jointly or separately, the US NASA and French CNES have played significant roles in advancing space technologies in support of climate observations, as exemplified by the US–French Topex/Poseidon oceanographic satellite (1992–2006), followed by Jason-1 (2001-present) and Jason-2/OSTM (to be launched in 2008). Another example is the so-called 'A-train' (see Figure 3.3), a constellation of satellites including the US platforms Aqua and Aura, the French Parasol microsatellite, the US-French Calipso (an innovative lidar mission), and the US-Canadian Cloudsat, all flying in formation in the same orbit at minute intervals to improve our understanding of the climatic role of clouds and aerosols.

Progress in surface-based observations

Surface-based observations also play a key role in climate monitoring. These cover a wide range of data acquisition systems located at the surface of the Earth (e.g. thermometers, barometers, anemometers, rain gauges, etc.), in the ocean (at the surface, such as tide gauges, fixed buoys, or drifters; or in-depth, such as moorings or 'pop-up' floats), or in the air (e.g. radiosondes, balloon-borne, or airborne sensors).

In recent years, funding reductions and other factors have led to unfortunate decline in the size and importance of surface-based observations. In fact, a stagnation or even decrease of *in situ* networks density, as observed in many countries, particularly in the Southern Hemisphere, may yet have a serious negative impact on our ability to assess climate and detect climate change.

Some significant advances have been made, however, including improved sensor quality and denser coverage of ocean measurements. The international Argo profiling float program is an example of a very successful endeavor, in which the United States and France play a significant role. Argo floats are oceanic robots that dive to a fixed 'parking depth' of about 2000 m and return to the surface every 10 days; they transmit via satellite their accurate measurements of temperature and salinity profiles acquired during the ascent. They then dive back for another cycle. Each float has a lifetime of 3–5 years, and 3000 such floats are planned to form a global array. Deployment started in 2000 and was complete in November 2007. Three float models are in use, namely the US Apex and Solo and the French Provor. Over 20 countries participate in Argo, with significant contributions of the United States (50%) and France (8%–10%) to the global array.

Argo floats complement the remotely sensed observations gathered by satellites and can be compared to the radiosondes, small balloon-borne instruments launched twice daily from 700 stations, which provide information on the vertical profiles of air temperature and moisture. It is worth noting that funding mechanisms for Argo differ widely between countries. Each nation has its own priorities, but all contributors subscribe to the goal of building the global array and to Argo's open data policy. Argo data centers are located in Monterey, California, and Brest, France, and data are freely available to any interested users through the Internet.

Global observing systems and international cooperation

Because the Earth is a complex, integrated system, the data needed to satisfy Earth system models must measure all climate-critical variables across all climate-relevant Earth subsystems. The priority need is not to measure every conceivable variable. In fact, the emphasis is on those variables that reduce climate model prediction uncertainty first and foremost and those metrics that we think are important to monitor, e.g. drought severity or changes in extreme precipitation. From the bottom of the oceans – across vast lands comprising deserts, glaciers, mountains, lakes, fields, and valleys – through the varied layers of our atmosphere – and on to the hostile, near-space environment – these essential measurements must take place on a continuous and coordinated basis.

This is an enormous undertaking – one that fully exceeds not only the financial abilities of any individual country, but also their geographic access. While one country might fly environmental satellites to routinely collect global surface and near-surface data from space, these data alone are not sufficient to fulfill all of the data needs. Surface and subsurface measurements – such as for soil moisture, land surface temperature, low-lying winds, and deep ocean currents – continue to require local (*in situ*) measurement in order to meet climate modeling requirements. Conversely, while these *in situ* measurements present the best data on local conditions at a given point in time, when taken alone they are insufficient to provide required climate information at global scale. Surface and subsurface *in situ* observing systems are unevenly distributed across land and ocean, with vast reaches of oceans and unpopulated land regions routinely accessible only by remotely sensed monitoring.

The components of a global observing system

In addition to the challenges of geographic scope, there are extensive challenges arising from the varying, original purposes of specific observing systems that also supply data for climate applications. Just as no one nation can reasonably provide a single, global observing system to meet all needs for climate data, a global observing system dedicated to climate is also an elusive dream.

Differing requirements, priorities, and budgets drive the capabilities of *in situ* and remote sensing systems. Short-duration research projects; long-term ('operational') regional and national weather prediction; wildland fire detection for fire fighting or biomass change detection; and sea ice monitoring for safety of navigation or local weather prediction: these are among the many and varied purposes of today's diverse observing systems. And, of course, the usefulness for climate monitoring of data that were originally intended for other purposes will depend on the compatibility of monitoring needs as well as on the degree of collaboration and cooperation during early planning for the particular system.

More complexity results from the differing data collection, processing, and storage schemes for those observing systems. Data from these systems have to be accessed in a variety of ways. Some are available only through direct collection from observing instruments at the time of data transmission or relay, i.e. in real time. Other data may be acquired from short-term data collection repositories. Environmental data archives, established to provide long-term access to observations, are also a critical data-access component.

The technical diversity of data dissemination paths and mechanisms for access to data from past, current, and future observing systems significantly increases the challenge of achieving a global climate observing system. Similarly, differing data policies add to the challenge. In order to gain financial and political support for the development of new observing systems, some nations are required to include elements of cost recovery in the business case for their system. In some other instances, system providers may be able to include free (or limited-cost) access to data – although perhaps on a delayed or lower data resolution basis.

Clearly, a global observing system for climate will result from the combination of diverse systems, both remotely sensed and *in situ*, used in conjunction with models, networks of data distribution, and ready access to the holdings of a wide array of environmental data-archiving centers. Given this reality, we can readily see the value of establishing broad agreements for international cooperation and collaboration across the many facets of realizing a climate global observing system capability.

The knowledge and evolution challenge

Although challenges abound in achieving a comprehensive and sustained climate observing system, the research community has succeeded in producing a continuing stream of innovation. Building on recent advances in satellite systems described above, promising new techniques are emerging, such as the use of infrared laser sources to probe the atmosphere and obtain vertical profiles of horizontal winds in the lower atmosphere (Aeolus of ESA) or vertical distribution of aerosols and clouds (Calipso of the United States and France); the use of interferometry techniques to derive continental soil moisture and sea surface salinity (SMOS of ESA and Aquarius of United States and Argentina) or forest biomass (ALOS of Japan); the use of high-resolution spectrometers to measure abundance of atmospheric carbon dioxide concentration, giving access to a description of carbon sources and sinks in the ocean and biosphere (OCO of the United States); the use of imaging altimeters to map polar ice sheets and derive the mass balance of ice caps (CryoSat of ESA); and the use of a constellation of satellites carrying passive microwave radiometer and a radar to determine global rainfall patterns (cooperative effort of the United States, Japan, and others).

All of these exciting new concepts have been developed by space research agencies, based on selected proposals received in response to regularly released announcements of opportunity. Even though most public research budgets devoted to Earth sciences are declining, in all countries there has been a steady flow of new ideas and new missions. Most have been highly successful, providing results much above expectations. Some can really be considered as 'one-shot' experiments, while many have the potential of becoming operational sources of climate data once demonstrated in flight – which drives us to the major challenges of transitioning observing systems from research to operational status.

The continuity challenge

As we work to achieve an effective global observing system for climate applications, efforts must focus on getting the right measurements together in the right form along with highly accurate climate models, which can then produce the information needed to make informed decisions related to climate, be it related to adaptation, mitigation, energy policy, etc. Yet just building the observing system is surely not enough. The observing system must be sustainable – producing accurate data on a continuing basis.

As the quality and availability of observing data and associated models improve, so will our understanding of the targeted Earth systems. This, in turn, will stimulate the need for further improvements in sensor technology, complex observing systems, and models. As shown above, such improvements are typically the product of research programs. Once their value has been established, these capabilities need to transition into ongoing, 'operational' capabilities.

Although recognition of this transition requirement is growing, building the migration path in the planning for current and future systems remains extremely difficult. Advanced techniques rarely flow quickly or smoothly from research into operations. The 'life cycle' of a complex, operational system may entail years for design and development, resulting in a series of system components (such as buoys, ground-based networks, or environmental satellites) that operate for decades, effectively locking in old technology and science understanding.

The funding needed to 'bridge' the results of a research program from one organization into the next-generation systems in an operational program of another organization is typically not within the budget of either institution. In fact, some of the more promising tools and techniques may prove entirely too expensive for long-term replication and operation. Some experts [138] have compared the challenge of bridging the gap between research and operations to 'crossing the Valley of Death.' Fortunately, organizations and governments are beginning to recognize and are attempting to address this challenge. The recent history of precision radar altimetry offers an interesting illustration of the kind of obstacles to overcome.

Starting in 1992 with the US–French Topex/Poseidon experimental mission, a 2.5-ton satellite carrying redundant instruments, the series evolved with the launch in 2001 of the French–US Jason-1, a 500-kg satellite with an optimized payload but equal performance (and a lifetime of 3–5 years, compared with the achieved 13.5 years of Topex/Poseidon). France and the United States essentially had an equal share on both missions. Responding to a letter of the international Science Team recommending pursuit of the series beyond Jason-1, the NASA associate administrator for earth science, G. Asrar, wrote in 1998:

It is essential to recognize that NASA has no mandate to continue such measurements indefinitely in the future. Doing so would jeopardize NASA's ability to pursue research and development into new areas. In order to acquire the necessary longtime series of global sea level observations, we need to work together to find ways of incorporating high-quality ocean altimetry measurements into future operational remote sensing programs.

This led to a four-party agreement on a Jason-2 mission between two space research agencies (NASA for the United States and CNES for France) and the respective operational agencies NOAA (United States) and Eumetsat (Europe). Reaching this agreement involved long negotiations. By that time, Jason-2 had to be largely redesigned due to satellite parts' obsolescence. The delay also meant that no operations overlap is expected between Jason-1 and Jason-2. The space research agencies have announced that they will cease their financial contributions after Jason-2, but neither operational agency has secured funding approval for a follow-up mission.

Achieving a global observing system for climate

Today, despite such difficulties, international cooperation and collaboration in observing systems take place on many levels and across a number of application domains. One of the longest and most extensive of these endeavors is the World Weather Watch (WWW) under WMO. The WWW combines observing systems, telecommunication facilities, and data processing and forecasting centers – owned and operated by WMO members – to make available meteorological and related geophysical information needed to provide efficient services in all countries. Although the WMO does not itself own or operate these capabilities, member countries have long recognized the mutual benefits possible from working together throughout the design, deployment, operation, and utilization of data from this network of observing systems. In many ways, the WWW serves as a role model for the results that are sought in the area of climate observations.

To address and engage the broader community (i.e. beyond weather services) in terms of both climate data requirements and contributing measurement systems, the GCOS program was established in 1992 to ensure that the observations and information needed to address climate-related issues are obtained and made available to all potential users. GCOS aims at establishing a long-term, user-driven operational system capable of providing the comprehensive observations required for monitoring the climate system, detecting and attributing climate change, assessing the impacts of climate variability and change, and supporting research toward improved understanding, modeling, and prediction of the climate system. Like WWW, GCOS itself neither makes observations nor generates data products. It provides an operational framework for integrating and enhancing, as needed, observational systems of participating countries and organizations into a comprehensive system focused on the requirements for climate issues. Although the scientific community has been strongly supportive of the GCOS activity in principle, progress in achieving a GCOS remains frustratingly slow.

The Committee on Earth Observation Satellites (CEOS) was formed in 1984 as an international mechanism to facilitate the coordination of international civil space-based missions designed to observe and study the planet Earth. In the 1990s, CEOS member agencies and associate organizations, including GCOS and others, recognized that in order to increase the effective utilization of space-based Earth observation data, a more thorough understanding was needed of major user requirements and how those requirements might be better supported by Earth observation data. Through a series of prototype projects, the concept emerged of establishing a broader collaboration among organizations involved with the planning and coordination of observing systems. Most importantly, it was recognized that for useful coordination to take place, users of these systems must also take part in the planning activities.

This led to the formation in 1998 of the Integrated Global Observing Strategy (IGOS) Partnership, which brought together activities that coordinate and promote the development of major Earth- and space-based systems for global environmental observations of the atmosphere, oceans (Figure 12.1), and land. The IGOS effort is widely recognized for its successful development of comprehensive, community-consensus 'theme' reports. These reports identify the Earth observation requirements of selected major application domains (e.g. ocean observations, geohazards, atmospheric chemistry, water cycle, carbon cycle) and present clear and compelling recommendations for action across sensor capabilities, system operational features, model enhancements, and data dissemination strategies.

Such cooperative efforts have advanced the understanding of what is required for successful global environmental observing systems in general, and for a climate observing system in particular. Yet in spite of such laudable efforts and the growing recognition of the vital role of Earth systems monitoring, much work remains to realize a comprehensive and sustainable climate observing system.

In recent years, several high-level events have for the first time brought political level attention to the need for progress in Earth observing systems. In the 2002 World Summit on Sustainable Development (WSSD) at Johannesburg, South Africa, governments formally recognized the urgent need for coordinated observations relating to the state of the Earth. Later, at a meeting of the Group of

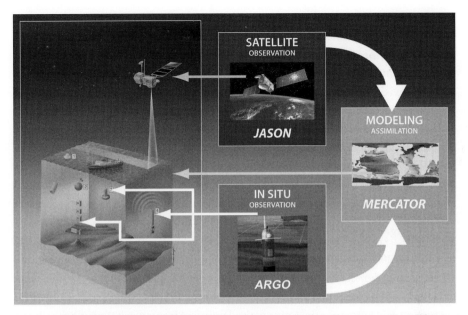

Figure 12.1 Artistic rendition of an integrated ocean observing system. (For image in color, please see Plate 15.)

8 Industrialized Countries (G8) Summit in June 2003 in Evian, France, the heads of state affirmed the importance of Earth observation as a priority activity.

In July 2003, the first Earth Observation Summit was convened in Washington, DC. There, governments adopted a declaration signifying *a political commitment* to move toward the development of a comprehensive, coordinated, and sustained Earth observation system of systems. The Summit established the ad hoc intergovernmental Group on Earth Observations (GEO) and tasked it with the development of an initial 10-Year Implementation Plan. After an intense period of activity over less than 2 years, the governments established a standing intergovernmental Group on Earth Observations (GEO) and an action plan to realize a Global Earth Observation System of Systems (GEOSS).

The vision for GEOSS is grand – to realize a future in which decisions and actions for the benefit of humankind are informed by coordinated, comprehensive and sustained Earth observations and information. GEOSS aspires to involve all countries of the world and to cover land and ocean observations as well as airborne and space-based observations. The established Earth observation systems provide essential building blocks for GEOSS. All Earth observing systems that participate in GEOSS retain their existing mandates and governance arrangements, supplemented by their contributions to GEOSS.

Probably the most significant promise of this initiative is that through GEOSS, participating systems will share observations and products with the system as a whole and take necessary steps to ensure that the shared observations and products are accessible, comparable, and understandable. In other words, 'the total will be more than the sum of the parts.'

In the climate domain, important goals for GEOSS include ensuring the sustained provision of data relevant to climate studies, promoting the enhancement of climate observing systems (especially in the terrestrial and ocean domains), improving access to quality-assured climate data, and assisting international coordination of climate observations. Rather than attempting to develop yet another plan for realizing a global climate observing system, GEOSS recognizes and supports the GCOS Implementation Plan.

Urgent action is needed

In 2004, GCOS published its Second Adequacy Report [139], in which the gaps and deficiencies that affect the present climate observing system were analyzed, and an Implementation Plan [140] providing a 10-year strategy for their correction. GEO provides the long-awaited political framework through which the demand to establish and maintain Earth observing systems comes from high levels within governments. However, without firm commitment and concerted action, there is still danger that existing systems will degrade in the coming years. Much has been achieved by current systems, but failure to take the opportunity afforded by GEOSS to rectify identified observation system deficiencies will mean that the potential to obtain substantial added value from the global observational network will be lost for the foreseeable future.

In certain important aspects (e.g. in surface climate, upper atmosphere, hydrological observations), the already insufficient observational capacity is likely to continue the decline that has been evident for several decades unless a decisive intervention is made. This is also true for satellites, which require long development times and involve substantial funding. Examples of forthcoming crucial data gaps or interruptions in the so-called 'Essential Climate Variables' are unfortunately numerous: as said above, there will be no overlap between the ongoing Jason-1 and its successor Jason-2, due for launch in 2008, and there is no plan for a Jason-3, which will mean the loss of continuity in global sea level measurements. There is a significant risk of lack of ocean color measurements as early as 2007 and almost certainly after 2009. Practically all ongoing experimental missions, however promising their results, remain without any plan for continuity.

With respect to new observation areas just emerging (e.g. around issues of energy and health), future coordination will be hampered by the failure to establish and adhere to interoperability standards at this stage. Others, such as aspects of climate change, land degradation and desertification, and biodiversity loss, failure to establish a comprehensive observation baseline at this time, and a commitment to continuity of observation systems, will hamper the ability to detect and quantify changes and the achievement of treaty targets.

Still, some positive signs should be noted, notably in the context of GEOSS. Prompted by a request of the UNFCCC Conference of the Parties and as part of its contribution to GEOSS climate needs, CEOS has developed in 2006 a plan to respond to GCOS requirements for space-based observations [141, 142]. In Europe, the *Global Monitoring for Environment and Security* (GMES) initiative launched in 1998 by the European Commission (EC) and European Space Agency (ESA) has moved from concept to reality, with a series of major funding initiatives since 2001. More recently, after an EC Communication affirming the intent to devote significant funding to the establishment and maintenance of an infrastructure for GMES as part of its contribution to GEOSS, the Ministerial Council of ESA held in Berlin in December 2005 took the decision to build a series of preoperational satellite missions called Sentinels. Initial funding by ESA member states will cover development costs, while the EC should support the continuation of the series. If confirmed, the Sentinels will ensure the continuing acquisition of many useful climate data sets.

The United States has also initiated a coordinated, national effort to achieve a US Integrated Earth Observation System (IEOS). The IEOS plan addresses several high-priority goals, including for the area of climate: 'Improving knowledge of Earth's past and present climate and environment, including its natural variability, and improving understanding of the causes of observed variability and changes.' The IEOS plan recognizes that climate observations need to be taken in ways that satisfy the climate monitoring principles (established by GCOS and formally endorsed by the UNFCCC), ensuring long-term continuity, and supporting the ability to detect small but 'persistent' signals. In addition, the plan highlights the need for an end-to-end Earth information system, including data management and data processing capabilities, as key to the success of the US effort in achieving an IEOS and supporting the GEOSS over the next 10 years.

The US and European efforts are promising. The GEOSS endeavor has to be a success story. History does not give second chances.

Key websites

GCOS	www.wmo.ch/web/gcos/gcoshome.html
GEO	www.earthobservations.org/index.html
IGOS	www.fao.org/gtos/igos/assets.asp
CEOS	www.ceos.org/
GOOS	www.ioc-goos.org/
GTOS	www.fao.org/gtos/

13

Climate and society: what is the human dimension?

ROBERTA BALSTAD AND JEAN-CHARLES HOURCADE

> The ultimate test of man's conscience may be his willingness to sacrifice something today for future generations whose words of thanks will not be heard.
> – Gaylord Nelson, former governor of Wisconsin, founder of Earth Day

Cet audivid alteram partem.
(J'ai écouté l'argument de l'autre.)

Introduction

Future historians of science will look back on the late twentieth century as a time when scientists made rapid strides in understanding the ways that the global environmental system, and more particularly, the global climate system, was changing. There was research taking place independently in a number of disciplines that focused on patterns and mechanisms of change in specific aspects of

---

Roberta Balstad, Columbia University, is a senior fellow at the Center for International Earth Science Information Network (CIESIN) and a principal investigator at the Center for Research on Environmental Decisions (CRED). She has published widely on the human dimensions of climate change. She was formerly President and CEO, and subsequently Director, of CIESIN and, prior to which, Director of the Division of Social and Economic Sciences at the US National Science Foundation.

Jean-Charles Hourcade, Centre National de la Recherche Scientifique and École des Hautes Études en Sciences Sociales, is Director of CIRED (Centre International de Recherche sur l'Environnement et du Développement). The author of many scientific papers in the field of development, energy, and environment economics, he was convening lead author for the second and third assessment reports of the IPCC and was a member of the French negotiating teams of the Kyoto Protocol, between COP-3 and COP-7.

Earth processes, including those of the oceans, the atmosphere, and the land. Over time, these separate elements were brought together to form a coherent picture of the interactions among the various parts of the global climate system. More importantly, scientists began to recognize the growing and cumulative role played by humans, through agriculture, technological innovation, and energy consumption, in causing these unprecedented changes in Earth's climate. To understand the interactions of climate and society, they turned with some trepidation to social and economic scientists for help. In this essay, we examine the ways that climate and society interact with and influence each other; how climate scientists and social scientists in France and the United States – and in other parts of the world – have been working together to understand the human dimensions of climate change; and finally, the ways that the growing scientific consensus on the significance of the human role in climate change is influencing national and international policies.

The Franco-American Consensus on Climate and Society

Scientists in France and the United States agree that climate and society interact in multiple ways. Social scientists, natural scientists, and physical scientists all conduct research that helps understand both the ways that human activity influences the climate (these are called anthropogenic influences on climate) and the ways that future climate impacts will influence individual and social well-being. For example, some climate changes are the result of changes in land use and land cover. Deforested land can reflect more sunlight back into space, and the extension of forests into areas formerly covered by tundra can increase the absorption of sunlight. The production of greenhouse gases traps heat in the atmosphere and warms Earth's surface. These greenhouse gases can be produced by both natural and anthropogenic processes, but in the past century, human activities have accelerated the production of these gases. Carbon dioxide (CO_2) has been increasing rapidly in the atmosphere because of the increasing use of fossil fuel sources. In a famous experiment, Charles Keeling began collecting data in 1957 on CO_2 concentrated in the atmosphere on Mauna Loa, a mountain in Hawaii. He found a steadily rising level of carbon dioxide year after year (see Figure 8.1 – the Mauna Loa curve). Another greenhouse gas is methane (CH_4), which is produced by livestock, the cultivation of rice in paddies, and landfills. All of these sources of methane involve human activities. Nitrous oxide (N_2O) is similarly produced by agricultural and industrial processes, and the combination of N_2O and hydrocarbons produces ozone (O_3). Several papers in this volume discuss the role of greenhouse gases (see, e.g. Chapter 2) and other physical processes leading to climate change in greater details. Here it is sufficient to point out that human actions, particularly those related to energy use and repeated over long periods of time, are instrumental in producing these gases and forcing many of these changes.

Both French and American scientists are also increasingly aware of the impacts that climate change will have on human societies. The warming of the climate will alter crop yield and increase the demand for water for irrigation. At the same time, the melting of polar ice caps could result in a rise in sea level on a global scale, leading to erosion of beaches and inundation of coastal areas. For low-lying and island communities, this could be devastating and result in the inhabitability of these lands, and for urban areas, many of which are in coastal locations, groundwater could be polluted. Ecosystems will change through loss or alteration of habitat, leading to movements of plants and animals to more benign environments and to invasions of new species. Both of these phenomena could lead to significant public health problems, population migration, and economic hardship.

It is difficult to predict how the climate will change in the future, but we know much more now about what causes such change, the nature of the human role in the process, and the potential consequences for human society. How we have learned about the interactions between climate and society is the focus of this paper.

Historical approaches to climate and society

Humans have always influenced their environment. In prehistoric times, human use of fire, the search for food, and the invention of agriculture all set in motion small changes in local environments that increased the dependence of human society on Earth's environment and altered the sensitive relationship between the environment and climate. As population size increased and humans began to use Earth's resources more intensively, both the immediate and the cumulative impact of human action began to play a larger, increasingly observable role in Earth's climate. With the advent of the industrial age in the nineteenth century, feeding, clothing, housing, and transporting the rapidly growing human population, especially in wealthy and industrialized countries, increasingly altered the relationship of human inhabitants with their planet and its climate. By the mid twentieth century, the explosive growth of the global population further intensified the use and transformation of Earth's resources and its climate (Figure 13.1).

The Norse settlement of Greenland during the medieval warming has long served as a cautionary tale of the impacts of climate on human societies. The mysterious decline and disappearance of the Norse settlements during the Little

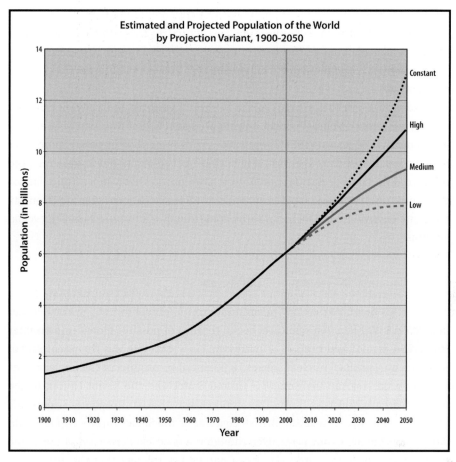

Figure 13.1 World population growth (source: UNESCO).

Ice Age, a few hundred years after they were settled, has recently been ascribed to gradual climate cooling that made it impossible for the Norse people to survive. There is some disagreement among the experts as to whether the fatal blow to the Norse settlements was delivered by the rapidly cooling European climate in the late Middle Ages or whether it was due to the Norse settlers' refusal to adopt indigenous food and hunting practices, which might have allowed them to survive. However, there is no disagreement that climate change was a critical element in the demise of the European settlements. Other research has been done on the role of drought in the decline and disappearance of the kingdom of Akkad in the ancient Near East.

Some historians have long recognized the role of human action in climate change. In France, Emmanuel Leroy Ladurie's *Histoire du Climat depuis l'an Mil* links cultural and institutional evolution to changes in the weather over a millennium.

He shows that environmental changes in the thirteenth and fourteenth centuries, with the greater incidence of famine and pestilential disease, were related to cycles of warm and cool climates. His work is important for its focus on climate as a force in human history and for illustrating the ties between environmental changes and social events.

In general, until the last half of the twentieth century, most people believed that technological progress was an unmitigated benefit and would help humanity overcome the constraints that climate and environment had imposed on pretechnological societies. They assumed, and both historical and social science research reinforced this interpretation, that nature had a progressively more limited influence on human activity. Gradually, with notable exceptions, any focus on climate and environmental factors disappeared from much twentieth-century social analysis.

Linking climate, science, and society

This technological optimism began to crumble, however, in the decades after World War II, when scientists began to warn of possible global catastrophes that could result from human activities. An early example was the possibility of what scientists called 'nuclear winter,' a potential by-product of the nuclear arms race between two superpowers, the United States and the Soviet Union, in the decades after World War II. Another concern was that stratospheric supersonic planes, such as the French and British *Concorde* or the Soviet *Tupolev TU-144*, could damage the Earth's ozone layer. Even though it was quickly proven that these concerns were scientifically unsound (see Chapter 11), the fact that the arguments were made showed the growing scientific and public awareness of the potential global environmental impacts of complex industrial products and processes. There was also a growing concern in the public about the environmental impacts of various types of ordinary human activities. In 1962, Rachel Carson published *Silent Spring*, a book that examined the environmental impacts of herbicides, particularly DDT, that were increasingly being used to improve agricultural productivity. Her book contributed to the growing concern that technological progress was not an unalloyed benefit and could actually be damaging to society.

Similarly, many people in the post–World War II world, both scientists and nonscientists, were concerned about the potential for disaster in the growing use of nuclear energy. In France, the government went on to build nuclear facilities that now provide about 80% of the country's electric energy, but in the United States, the construction of nuclear facilities came to a stop after the accident at Three Mile Island in 1979. These issues – nuclear energy, deadly herbicides, damage to the ozone layer, acid rain, and nuclear fallout – cast scientists in the role of linking the impacts of industrial production, energy consumption, agriculture, and the public welfare in new ways and illustrating the potential for social and political controversies that could result from scientific research.

Even as the public was beginning to be concerned about what scientists were saying in relation to the human role in the environment, climate scientists were beginning to examine what appeared to be gradual but significant changes in Earth's climate. Scientific research on climate change was not initially driven by a concern about the role of human activities on the climate. Climate research had traditionally been conducted by physical scientists who were largely unequipped to conduct research on the human component in the geophysical phenomena they were examining and who often considered socioeconomic research outside the scope of their research tools. Their pathbreaking research exposed the workings of the climate system and the nature of climate changes on a global scale, but it described the physical, not the human, dynamics behind the changes they observed. Some of the early work by Wallace Broecker [42] on abrupt change in deep ocean currents emphasized the global, physical processes related to climate change. Other research in the 1970s showed how industrially produced chemicals were altering the composition of the Earth's atmosphere through the release of chlorofluorocarbons and the consequent depletion of stratospheric ozone [143, 144, 145]. By the late 1980s, there was increasingly broad recognition among scientists that the Earth's atmosphere and climate had been undergoing unusual and possibly unprecedented changes on a global scale since the Industrial Revolution, and that the pace of these changes was accelerating.

Since the 1920s, however, social scientists in both countries had been technological optimists, focusing their research on the social and technological worlds, including the study of social, political, and economic forces in society and the social, cultural, and individual influences on human behavior, instead of on the physical world. In the United States in the nineteenth century, some social scientists were interested in the ways humans were altering their environment (e.g. George Perkins Marsh and the study of anthropogeography). But over the course of the twentieth century, a century characterized by the use of technology to overcome the limitations that the physical world had previously placed on the social world, research in the United States on the influence of climate and environment on human and social behavior gradually dried up. Stimulated by the energy crisis of the 1970s, economists became concerned about the ways that humans were consuming their natural resources, and some social scientists like the geographer Gilbert White and his students continued to emphasize how the human population was changing the natural environment. In the United States, though, social scientists generally studied society, not the physical environment within which society functioned. Similarly, in France, philosophers such as Théodore Monod and

Jacques Ellul began to warn of potential environmental problems resulting from technological change, but most social scientists tended to ignore this and focused their attention instead on equity issues and the social costs of economic growth. In both countries, the report of the Club of Rome in 1972, with its emphasis on the natural limits to growth and development, challenged the technological optimism characteristic of so many people in the postwar period. Equally important in this challenge was the work of developmental economists such as Gunnar Myrdal and, in France, René Dumont on the detrimental impacts of economic development on local ecosystems in Africa.

In short, when climate scientists turned to social scientists for assistance in understanding the anthropogenic implications of climate change, few social scientists were prepared to undertake research on climate and society or even interested in collaborating with climate scientists. Studies of human influences on the natural environment and its resources were not the same as research on the human role in climate change and did not immediately stimulate such research. There were a number of reasons for this. In part, this was because the 'environment' is usually interpreted in local rather than global terms. Whether a locality was defined as an ecosystem or an administrative jurisdiction (that is, an area bounded or defined in political or historical terms), research on how humans were influencing their environment, like most social science research, was focused on specifically defined localities. Climate, however, is a global, not merely a local, phenomenon.

Involving social scientists in climate research

Another reason that social scientists were unprepared to undertake research in partnership with climate scientists is that they knew little about climate science, just as the climate scientists knew little about social science. They were not familiar with basic concepts of climate science or with the research questions or methods of climate scientists. Gradually, however, social scientists began to examine issues that are related to both the impacts and the causes of climate change. For example, some individuals working at the interface between economics and engineering became interested in the interface between energy and the economy. These analysts were used to dealing with decadal time periods because of the long turnover of investment in the field of energy and research on long-term safety of nuclear energy. At the same time, growth theoreticians such as Robert Socolow and William Nordhaus had responded to the issue of the 'limits to growth' through the development of models of growth encompassing more than a century. In France, this work was also being fostered by development economists such as Ignacy Sachs under the concept of ecodevelopment, a precursor of the idea of sustainable development popularized by the Bruntland Report on environment and development in 1986. For these specialists, climate issues were a stimulus to integrate several research and policy concerns, including energy efficiency, consumption and technological development, energy security, long-term impacts of current behaviors, and the reconciliation of environmental protection and the alleviation of poverty. Many of these issues are still being examined today.

At the request of governments and nongovernmental organizations (NGOs), energy modelers became involved in early attempts to explore the causes of anthropogenic climate change and produce sets of long-term emissions scenarios corresponding to various views of the development of the global economy. These scenarios also took into account anticipated changes in demography, economic growth, and technological change. In response to questions from the policy debates, this group began to work in synergy with other scientists and engineers. A difficult but very fruitful dialogue took place between economists and engineers about the capacity of current technologies and of technological innovation to support ambitious policy goals of 'decarbonization' of the economy.

In both France and the United States, the main controversy was between the optimism of the engineers and the pessimism of the economists. The latter grew out of concern about the need for viable policy instruments for mobilizing new technological options (such as standards, carbon taxes, and tradable permits) and the potential for negative impacts of these options on growth, income distribution, and industrial competition. The research communities in the two countries have worked closely together at least since the end of the 1980s. Since then, it has been difficult to distinguish between French and American positions – to say, for example, that French economists favored command and control policies and US economists favored tradable permits in a pure market economy. Early on, economists of both countries underscored the limits of both command and control approaches and the use of economic instruments. In both countries, some economists advocated carbon taxes to reduce distortions in the tax system and produce a double dividend, and in both countries the same diagnosis was made about their potential political problems. Similarly, there was no national consensus among economists about the value of worldwide carbon trading or how to solve the problem of the allocation of emissions allowances. Perhaps the most significant difference was that French economists tended to emphasize economic development goals for the Third World more than their US counterparts did.

When social scientists outside economics were first attracted to research on climate change, they continued to ask traditional social science research questions but focused them on climate issues. For example, social scientists were interested in such questions as how societies manage common pool or jointly held resources, like water or grazing commons, what people think about the environment, and how their consumption patterns affect natural resources. There continues to be a heavy emphasis on the role of energy and production systems, the use of fossil fuels, and the production of greenhouse gases in changing the climate. More recently, social scientists have begun to focus their attention on such research issues as the ways that societies understand and communicate information about climate, how environmental policy decisions are made, especially under conditions of climate uncertainty, and how to change behaviors that are environmentally destructive. A key challenge is to understand what influences the development and implementation of new, climate-friendly policies.

Although there has been a growing scientific consensus about the significant role of human action in climate change, political understanding of that role and attempts to moderate it often lag behind the scientific understanding. In the United States, public understanding of the impacts of climate change was advanced through a congressionally mandated national assessment of the impact of climate on various regions throughout the country. Begun in the late 1990s, the National Assessment built on what was known about the local impacts of global climate change and tried to determine what this meant for specific regions of the United States. It consisted of multiple studies that were conducted by partnerships of local scientists and non-scientifically trained stakeholders in order to combine the knowledge of the scientists with the socioeconomic priorities of the stakeholders. These reports brought the impacts of climate change and climate variability to the attention of many policy makers throughout the country.

They also stimulated the interest of social scientists in integrated assessments, combining regional modeling efforts with case studies to improve our knowledge of climate impacts on local scales, particularly in fragile developing economies. Both US and French economists have attended the same meetings on integrated assessment, including an international workshop in Toulouse co-organized by one of the authors of this paper in 1996 titled, "Prospects for Integrated Environmental Assessment: Lessons Learned from the Case of Climate Change?" The fact that scientists had paid little attention to local impacts of climate change prior to this was due in part to our ignorance of how much the uncertainty of large climate models is exacerbated at the local scale. Yet the multiplication of extreme climate events is changing the political dialogue and will probably stimulate more research on local risks and impacts of climate variability and change.

National and international policy responses

Climate change is fostering changes in national and international policies and economies. One result of the growing international recognition that human actions are driving climate change is the attempt by many governments to develop local, national, and international policies to reduce or control human damages to the global climate system. National policies are obviously influenced both by external and internal pressures. That is, national policies are shaped in part by a country's commitment to international agreements and in part by political priorities and pressures within the country. This includes pressures by non-governmental organizations, 'green' parties, scientists, and concerned citizens, and, from the opposite side, it includes economic and political pressures exerted by firms and corporations that are reluctant to change the current state of affairs and the lifestyles they make possible under the presumption of a nontangible risk.

Scientists have attempted to minimize the communications gap between themselves and policy makers in several ways. One successful example is the ongoing interaction between scientists and diplomats during the negotiations that led to the adoption of the Montreal Protocol in the 1980s [146]. The scientists organized independent and parallel meetings to identify the issues for which there was scientific consensus regarding the role of chlorofluorocarbons in the stratospheric ozone layer. The Montreal Protocol is discussed in detail elsewhere in this book. Here, however, it is useful to point out that scientists effectively communicated with diplomats and politicians about the nature of the scientific consensus and issues for which there was no consensus. They believed that it was critical that diplomats not be confused or misled by claims of scientific uncertainty where little uncertainty existed. The conclusions of the scientists were often conveyed to the diplomats and politicians by scientists from their own countries who had attended the meetings. This involvement of scientists from around the world added to the confidence that negotiators could place in the scientific conclusions. In short, the scientists' attempts to develop a consensus during the process leading up to the adoption of the Montreal Protocol contributed to a more effective role for science in the negotiations.

The Intergovernmental Panel on Climate Change (IPCC) has played a similar role in translating scientific information to policy makers in both government and business by identifying areas of scientific consensus and preparing reports for non-scientists. Initially organized by the G7 in 1988, the IPCC is discussed in Chapter 1, but it is important to emphasize here the role of the IPCC not only in the gradual realization that climate change will have an impact on Earth's human population, but also in recognizing that human actions are ultimately responsible for current climate change. The IPCC had a broader scientific agenda than the meetings attended by scientists in the 1980s in parallel to the diplomatic meetings on policies related to stratospheric ozone depletion. In IPCC deliberations, the entire chain of energy extraction, production, and consumption was discussed, and there was never any question that policy responses to climate change were necessary.

The participation of social scientists in this process increased significantly over the course of successive IPCC reports, which came out roughly every five years. Of the three working groups established for the first IPCC report, one dealt with the science of climate change, one with socioeconomic impacts of climate, and the third with response strategies. Similarly, a decade later in 2001, the IPCC focus was on three topics: the science of climate change; impacts, adaptation, and vulnerability resulting from climate change; and mitigation of climate change.

The idea behind the formation of the IPCC is that policy makers in both government and the private sector need to know about recent scientific research findings on climate change and broad areas of scientific consensus from a source they can trust. The emphasis in IPCC reports is on comprehensive, objective, open, and transparent discussions of human-induced climate change and its potential impacts, and the IPCC process includes the participation of scientists from countries around the world. Moreover, because climate change is a field in which there is a great deal of research under way, the IPCC has been conducting a new scientific assessment every five years.

The role of social scientists was critical to the IPCC deliberative process in several regards, including the development of future scenarios, debates about the timing of responses to climate change (e.g. whether to act now or later), the costs of various abatement objectives, and the types of instruments most likely to minimize the cost of remediation. These matters can be both controversial and politically sensitive because they are defined by individual perceptions of risk, various degrees of optimism about the capacity of society to respond to policy signals, and diverse conclusions about the acceptability and performance of various policy approaches. The goal of the work of social scientists has not been to identify the single approach that is most accurate from a scientific perspective, but rather to rationalize the discussion by disentangling the sources of misunderstanding from the real and legitimate points of contention. Although there were differences of approach among US and French social scientists, these were similar to differences among US social scientists or French social scientists. They were not, in short, determined by nationality. For example, early on, economists in both France and the United States agreed about the limits of command and control approaches. Moreover, in both countries, carbon taxes were advocated to reduce the impact of distortionary taxation, but social scientists recognized that this approach was politically difficult in both countries. Possibly the most significant difference is that French social scientists in the past tended to give greater emphasis to the links between environment and development, although in recent years, this approach has been adopted by many social scientists in the United States as well.

The connections made by social scientists in the IPCC process were useful in discussions leading up to the Kyoto Treaty. Their links to social scientists in other countries and to decision makers in their own governments served them well in their attempts to promote a common understanding about the consequences and unintended impacts of the climate policies under discussion. Part of this effort was devoted to assessing costs of meeting Kyoto targets (including various forms of the double dividend) and another part was about the jurisdictional implications of the protocol and its adaptability to various institutional contexts. This was only a semisuccessful process, since the Kyoto Protocol remains an 'unfinished business.' However, it demonstrates the capacity of social scientists, along with others, to contribute to the emergence of a common language to deal with climate change policy.

It is clear that the Kyoto Treaty, which has been adopted by most nations, is having an impact in both France and the United States. In France, public opinion leads most politicians, whatever their political position, to claim support for climate policies. This has led to a series of reports from the Senate and National Assembly and to the adoption of a national *Climate Plan*, which includes an ensemble of measures for supporting energy efficiency and technical innovation. France's allegedly 'virtuous' position is facilitated by its extensive use of nuclear power, a clean source of energy, and its limitations on industrial emissions. Yet the country may find that future cuts in emissions, required by Kyoto, will be difficult until it has developed plans for greater energy efficiencies in the three principal energy-consuming sectors: housing, industry, and transportation. These will be influenced by both European policies, such as the European Carbon Trading System, and national politics.

In the United States, various policy arguments have been advanced to explain the US decision not to sign the Treaty. These include the observation that because the emphasis in the Kyoto Treaty is on current rather than future emissions, it represents little more than a partial and short-term strategy. Although the United States contributes significantly to global emissions today, its global share will drop considerably by 2030 because of the rise of emissions in China and India. Action taken to reduce the US share today, according to this argument, will be less important in both the short and long term than actions to reduce future rapid growth in emissions elsewhere. The United States also places a strong emphasis on the role of technology transfer in reducing future emissions in the developing world and argues that this is a more cost-effective approach to emissions reduction globally.

Although the US government is not a signatory to the treaty, the treaty is still having an impact in the United States. US corporations that have maintained operations in countries that have signed on to the treaty are forced to observe the Kyoto provisions in those countries. In addition, there are hundreds of local governments in the United States that have voluntarily agreed to adhere to the Kyoto standards and at least 10 states that are attempting to reduce emissions in line with the Kyoto Treaty. Equally important, acceptance of Kyoto and concern about emissions is not necessarily related to political party allegiance. For example, two very large states, New York and California, with a Republican governor, have assumed regional leadership in attempts to limit greenhouse gas emissions on a statewide basis. In addition, corporate and religious leaders in the United States are increasingly expressing concern about the impact of human activities and emissions on the global climate and seeking to limit these impacts. In short, although France and the United States have taken different political paths in regard to adherence to the Kyoto Treaty, the concern about emissions is widespread in both countries.

Conclusion: climate is a social issue

Although much of the early scientific research on climate and climate change was initiated and conducted by natural and physical scientists, it is clear that climate is a social issue in both France and the United States and that it must be addressed by social scientists as well as other scientists. Society is directly influenced by climate and climate variability in ways that we are only beginning to understand. Equally important, social policies and practices, and the actions and decisions by individuals and social groups repeated over time and space, have and will continue to have a crucial effect on Earth's climate. In this context, climate and climate variability has become an element that influences the plans and policies of nations and the hopes and aspirations of individuals. It has the potential to alter the 'reality' that we have become accustomed to in the past and to introduce changes not only in the physical world we inhabit, but also in our social and economic organization and use of that world. It links all the peoples of the Earth in a common endeavor and forces us to consider responses to the climate system as a shared responsibility. French and American social scientists must continue to work side by side with other scientists and policy makers from both the developed and the developing world in this common endeavor.

Conclusions

S. ICHTIAQUE RASOOL AND JEAN-CLAUDE DUPLESSY

Humans do not tread softly on the earth.
– National Academy of Sciences, *One Earth, One Future*, Washington DC, 1990

Nul remède précipité ne peut suppléer à un arrangement fixe et stable, établi de longue main, et qui pourvoit de loin aux besoin imprévus.
– Voltaire, *Le Siècle de Louis XIV*, Chap. XXX, 1751

The evolution of the twentieth century

All throughout the preceding chapters, the authors have emphasized that the climate in the twentieth century has been so abnormal that there was no precedence of it in recorded history. Climatic events that were often called 'once in a century events' have been more 'frequent,' especially in the last two decades, and records of all kinds (e.g. temperature, precipitation) have been broken several

---

S. Ichtiaque Rasool is a former Chief Scientist for Global Change at NASA and is now an independent scholar in the area of remote sensing of parameters related to carbon and water cycle research. Until 2001, he also occupied the position of Senior Research Scientist at NASA's Jet Propulsion Laboratory/Caltech, Pasadena, California, and was a visiting professor at the Complex Systems Research Centre of the University of New Hampshire in Durham. From 1990 to 1997, he directed the IGBP-DIS program from the University of Paris, France. Dr Rasool has received a number of awards from NASA, including the highest scientific award, the medal for Exceptional Scientific Achievement, in 1974. He received the William T. Pecora Award in 2002 'on outstanding and sustained international leadership in advancing remote sensing as a fundamental element in Earth System Science.' Dr Rasool was born in India in 1933. He earned his doctorate in Atmospheric Sciences in 1956 at the University of Paris. In 1970, he became a US citizen and currently lives in Paris, France.

Jean-Claude Duplessy is Director of Research at CNRS (the French agency for fundamental research), which he joined in 1967. His research activity deals with past climatic variations and their implications for the climate of the next centuries. From 1985 to 1996, he directed the Centre des Faibles Radioactivités (a joint laboratory of CNRS and CEA) and chaired the French "Global Change" Programme.

times. Excessive rainfall followed by massive flooding occurred in one region of the world, while simultaneously, in another part of the globe, intense and persistent drought resulted in huge losses of agriculture and life. The rapid rise in global mean temperatures by 0.8 °C (1.4 °F) in the last 100 years, and noticeably by 0.6 °C (1.1 °F) during the last three decades, is certainly unprecedented when compared with the records of the last thousand years, during which there is no evidence of a warming trend as global, synchronous, and persistent as that observed in the last few decades. Regionally there are many other examples of 'extreme' events, such as very high temperatures in Alaska and Siberia, the intensification of the hurricanes in the equatorial Atlantic, the heat wave over Western Europe in 2003, snow-cover decrease over land, and the shrinking ice sheet over Greenland, among others.

Although scientifically extremely intriguing, these climatic changes are often lumped together as 'global warming,' but we have yet to find a single 'smoking gun' for this warming scenario, if not the rather large influx of CO_2 in the atmosphere over the last century. In fact, the natural climate system is so complex and has so many feedbacks, both positive and negative, that 'warming' and 'cooling' scenarios, at varying time- and space scales are always at work, thus making prediction of the trend in global warming for the next several decades highly tentative. Greenhouse gas emissions, clouds, volcanic emissions, aerosols, oceanic circulation, melting of ice sheets, land surface changes, and above all, the population growth of about 80 million people per year all affect the regional and global climates and their impact on society. In what follows, we attempt to dissect the problem, prioritize the issues, and present some diagnostic solutions to reduce the uncertainties in the future climate scenarios, which we all will be facing together.

Perturbation of the composition of the atmosphere by human activities

Atmospheric content of CO_2 has continuously been increasing since the beginning of the pre-industrial era. Its concentration was about 280 ppm (i.e. 280 cm^3 CO_2/m^3 air) in 1750 and is near 380 ppm in 2006, of which about two-thirds are from fossil fuel burning and the remaining one-third is from land-use activities, such as forest clearing, agriculture, etc. We know for sure that most of the CO_2 increase is due to human activities (see Chapter 8 on carbon cycle), because we also have continuous measurements showing that the atmospheric oxygen (O_2) content decreases as a result of fuel combustion. Isotopic markers (C^{13}/C^{12}) confirm independently that the CO_2 increase originates from fossil fuel combustion.

Global mean CH_4 concentration increased from about 750 ppb (i.e. 750 mm^3 CH_4/m^3 air) in 1750 to more than double, 1775 ppb, in 2006. However, it has not increased any more since 1999, and this new trend is poorly understood. The magnitude of sources and sinks is also not very well-known, either. Most of the methane in the atmosphere is of biospheric origin, like wetlands, cattle, rice, etc., but how much is from each of these sources is another matter. Another 10%–20% comes from production and use of fossil fuels. The removal of CH_4 in the atmosphere is mostly due to its interaction with atmospheric hydroxyl radical (OH), and therefore its lifetime is limited, about 12–14 years.

N_2O, also a greenhouse gas, continues to increase in the atmosphere at a rather slower rate but because its lifetime in the atmosphere is long (about 120 years), it has a potential impact on the climate system.

It is clear that human-made greenhouse gases globally continue to increase. If the Montreal Protocol made it possible for chlorofluorocarbons (CFCs) to begin to decline, they were replaced by other gases, which are less aggressive to stratospheric ozone but have strong greenhouse efficiency. Their concentration is also rapidly increasing. Because of the partial ban on CFCs, stratospheric ozone is seemingly beginning to stabilize. However, the decrease in ozone concentration above 12 km results in stratospheric cooling. Tropospheric ozone, on the other hand, continues to increase specially in the Northern Hemisphere as a consequence of regional pollution and hence has a warming impact on local and regional climate.

Aerosols resulting from human activities have also been increasing as continuous measurements, either *in situ* or by satellite and land-based remote sensing instruments indicate. Their role, both direct and indirect (modification of cloud types, height, etc.), is mainly to produce cooling at the surface.

Perturbation of land and ocean

Land surface cover is very different from that which may be expected from natural vegetation. Forests have been cut to provide construction wood, croplands, and pasture. These activities have strongly increased since 1700 AD. The rate of forest clearing worldwide now totals approximately 1.4 million hectares per year, about half the size of a country like Belgium. The repercussions of such a massive change in land cover are enormous and will be discussed below.

The ocean receives the soluble wastes released by human activities, noticeably nutriments. It also absorbs part of the CO_2 injected in the atmosphere, and the oceanic CO_2 concentration increase as a consequence of human activities is now detectable. We are beginning to observe the changing acidity of the oceanic regions, where CO_2 penetration is concentrated (e.g. North Sea), and its detrimental effect on marine life.

Physical consequences

Due to their greenhouse effect, the long-lived gases CO_2, CH_4, N_2O, halocarbons, and others such as sulfur fluoride (SF_6) increase the radiative forcing, i.e. the radiative imbalance at the top of the troposphere (see Chapter 2). A positive radiative forcing would result in a warming of the lower atmosphere, whereas a negative forcing would lead to a cooling. Since the preindustrial era, the present-day radiative forcing increase due to increased concentration of greenhouse gases is about 2.6 ± 0.25 W/m^2 and is continuously increasing. Tropospheric ozone also contributes to warm the troposphere, although regionally, whereas the stratospheric ozone decrease results in a measurable global stratospheric cooling.

Aerosols scatter the solar radiation and some sulfate aerosols also act as nuclei to form cloud droplets, thus modifying the cloud cover. This has a negative radiative forcing (cooling) at the surface. On the other hand, charcoal/soot aerosols in the lower atmosphere may have a warming effect, but over a limited region of the Earth. Land-cover changes (deforestation, agricultural land use) have increased the surface albedo, whereas the black carbon particles resulting from biomass burning when deposited over snow and ice decrease albedo and have a positive radiative forcing.

Altogether, we are sure that human activities have modified the climate system (atmosphere, oceans, and land surfaces), and it is highly probable that these changes produce a net warming effect on the Earth's climate. The magnitude of this warming strongly depends on feedbacks and interactions within the climate system (for instance, changes in the cloud cover, continental albedo, sea ice extension, ocean circulation, etc.).

Evolution of the air temperature

Global mean temperatures averaged over land and oceans have risen by about 0.8 °C (1.4 °F) since the beginning of the twentieth century and by 0.6 °C (1.1 °F) over the last 30 years. The warming trend has been accelerating since about 1980. The years 1998 (a strong El Niño year) and 2005 have been the warmest years since meteorological observations have been available.

Warming has occurred over both land and ocean, but land regions are warming faster than the ocean. This warming is associated with a reduction in the number of frost days and cold extremes in mid-latitudes, whereas the number of hot days increases.

Other changes within the climate system

There is now strong observational evidence that the global ocean, even at great depth, has experienced a detectable warming recognized by compilations of temperature data collected since the beginning of oceanographic expeditions.

According to observations gathered by Bryden et al. [147], the mean heat flow carried by warm waters from the equatorial Atlantic to the north seems to have been slowing over the last 50 years. The decadal variability of the thermohaline circulation is at present too poorly known to consider this reduction as a stable trend, but we should clearly pay attention to ocean circulation changes during the next years and decades.

The hydrological cycle is changing, although patterns of precipitation are spatially and seasonally variable. A significant precipitation increase (0.5%–1% per decade) over the continents north of 30 °N has been observed since the beginning of the twentieth century. The average annual discharge of freshwater from the major Eurasian rivers to the Arctic Ocean has increased significantly over the last 60 years, and heavy precipitation events have increased over the last decades.

At the same time, droughts have become widespread in many places of the world over the last 30 years, which have a negative impact on human life, especially in regions where the rate of population increase is about 2–3 times larger than the global average.

The atmospheric water vapor content has increased, as might be expected from a surface temperature increase. As water vapor is also a greenhouse gas, a good knowledge of its increase in the atmosphere is critical for climatic model projections of the future.

El Niño events in the Eastern Pacific have a worldwide effect on global climate, as does La Niña. In recorded history, they have occurred regularly at a rhythm of about once in a decade. Now it is becoming more frequent, with a periodicity of 2–7 years, resulting in more frequent atmospheric circulation changes. Snow and ice cover observed by satellites has decreased over the last 30 years with the Arctic sea ice decline both in thickness and extent; the Greenland ice sheet is significantly shrinking, resulting in a sea level increase of about 0.2 mm per year (with the likelihood of a more rapid increase in the future due to nonlinear effects), and the tropical mountain glaciers are rapidly retreating.

Lastly, the sea level has increased during the twentieth century. Recent estimates range between 15–20 cm for the whole century. Direct measurements since 1990 by satellites indicate an increase of the rate of sea level rise, which at present is 3.1 ± 0.8 millimeter per year.

Lessons from the past

The climate has permanently changed during the Earth's geological history (of about 4.5 billion years). The best-known period is the recent Quaternary. Over the past 800 000 years, both marine sediment and ice core records show that the climate alternated between warm periods more or less similar to the present and glacial periods characterized by colder temperatures and the waxing of continental ice sheets. Over this entire period, a close correlation is observed between air temperature and atmospheric CO_2 content.

The climate is highly sensitive to minor perturbations of the radiative budget of our planet. For example, the Earth's orbit around the Sun is slowly changing, because not only the Sun but also the other planets of the solar system attract the Earth. Changes in the orbital parameters occur with periodicities of 21 000 and 41 000 – about 100 000 and 400 000 years. They lead primarily to changes in the intensity of the seasonal cycle, whereas the mean annual amount of heat received from the Sun by the planet Earth is almost constant. The effect of these seasonal variations is tremendous, and they act as a 'pacemaker' of Quaternary climatic variations. The astronomical theory to explain these oscillations was first presented by Milankovitch and assumes that small changes in the Earth's orbital geometry are the fundamental causes of the succession of the Quaternary ice ages. It is now widely accepted.

Superimposed over these long-term climatic variations, both ice and marine climatic records also show that large (more than 10 °C or 18 °F), widespread, abrupt (over less than 100 years) climate changes have occurred repeatedly throughout the last glaciation. Major cooling events were related to periods of massive iceberg discharges, which have punctuated about every 7000 years or so over the last glacial period. It began 115 000 years before present (BP) and culminated about 20 000 years BP. A very rapid warming over the whole North Atlantic area marks the end of the last major iceberg discharge, about 11 000 years BP. These changes are explained by an abrupt reorganization of the climate system: iceberg discharges resulted in a massive injection of freshwater in the high latitudes of the North Atlantic Ocean, which slowed down the thermohaline circulation and the oceanic heat transport to the Northern Hemisphere high latitudes. When the freshwater injection due to iceberg discharge into the North Atlantic stopped, the oceanic circulation and the associated oceanic heat transport started again. Such abrupt climatic changes have occurred only when ice sheets covered the high latitudes of the Northern Hemisphere continents. Climatic conditions were therefore significantly different from the present, but the occurrence of these abrupt changes demonstrates that on the decadal timescale, the North Atlantic freshwater budget and circulation is a critical parameter of the Earth's climate.

The twentieth century in the context of the last millennium

Tree rings, mountain glaciers, ice cores, lake, and marine sediments together with historical data form the climate records of the last millennium. These records have provided evidence for a medieval warm period followed by a climatic deterioration called 'Little Ice Age' in the sixteenth–nineteenth centuries. A large spatial and temporal variability is observed during the last millennium. Warm years, sometimes locally warmer than recent years, were recorded, but there is no evidence during the last millennium of a warming trend as global, synchronous, and persistent as that observed during the twentieth century. It is thus clear that although Earth's climate has gone through substantial natural changes in the recorded history, the warming events of the last 30–40 years are so strong and so rapid that they are highly likely due to anthropogenic sources.

Living in a perturbed world: follow the well-proven 'Genesis' strategy!

It is clear from all the chapters in this book that for a given emission scenario, we are yet unable to accurately forecast changes in most components of the climate system. Also, it is obvious that today we cannot reliably estimate changes at the regional scale that have the greatest societal interest.

The reason for these uncertainties stem from the fact that our climate model projections over long periods (decades to centuries) do not yet correctly replicate, for example, the physics of cloud-radiation processes, the role of aerosols in perturbing regional and global climate, and the dynamics of thermohaline oceanic circulation; and finally, they do not provide proper linkages between meso-scale weather and the large-scale general circulation of the atmosphere.

In order to rectify the current situation of uncertainty in predicting, for example, how much the temperature will increase and/or whether there will be droughts or floods in a given region in the next decades, we propose a five-prong approach of a scientific 'genesis' strategy, which means, 'Let there be light' on the following topics and take some critical actions immediately.

1. Understanding 'critical' aspects of the climate system: regional versus global

Each of the preceding chapters has highlighted a number of outstanding unsolved issues that require immediate attention. In the following, we describe a few of them, which in our judgment fit the genesis strategy:

1. As stated in Chapter 4 by Chahine and Morel: "There is really no alternative to developing much more realistic numerical representations of real-life moist atmospheric processes at their proper scales, and testing these models against detailed observations of the time-dependent three-dimensional structure of precipitating clouds. Such models would be instrumental in establishing the linkages between meso-scale weather and cloud-scale physics, on the one hand, and the large-scale general

circulation of the atmosphere and climate on the other hand. [...] It is time finally for climate science to reconnect with meteorological science."

2. Atmospheric aerosols present a rather challenging problem to those scientists studying climate change. First, their impact on radiative forcing could be positive or negative depending on their nature, sizes, origin, and whether they are hygroscopic or not, and of course their spatial and temporal distribution. Aerosols resulting from human activities are superimposed over those continuously produced by natural processes. Globally, on the other hand, SO_2 emanating from strong volcanoes, e.g. Mount Pinatubo, may reach stratospheric levels, react with water to become sulfate droplets, and circle the planet for years in a row, cooling the global temperatures by about 2 °C (3.6 °F) until the stratosphere is essentially clean (volcanic aerosol-free) again in about 2–3 years! It is obviously very difficult to predict all these man-made and natural events, which have a short duration and no periodicity, but a strategy can and should be developed to monitor aerosol levels, their nature, and large events both regionally and globally.

3. Wunsch and Minster underline in Chapter 6 the importance of an ocean observation program that has an eye on "slow, steady, nondramatic shifts" in oceanic parameters and is maintained for decades to come. This way, the scientists and the coastal population are aware of the consequences of, for instance, a rise in sea level, because the effect on the oceans of global warming will be slow to come but the dangers are increasing steadily with time.

4. Moore and Ciais emphasize in Chapter 8 that, in addition to the uncertainties on the global carbon cycle, the 'estimation of *regional* carbon fluxes is a significantly underdetermined problem. For some regions like North America or the vast expanse of Eurasia, we do not know even if the region is acting as a net carbon source or a net carbon sink from year to year. Simply stated, the current set of direct *in situ* observations is far too sparse, and this sparseness hinders geographically resolved source-sink determination.'

Finally, we should never forget that the evolution of the climate perturbed by greenhouse gases and dust emitted by human activities will be superimposed over the natural evolution of the Earth's climate. Natural variability of the climate system is still poorly understood but remained within limits that paleoclimatology allows to describe. Most components of the climate system exhibit natural, yet unexplained, variations with various time constants, some rather short for the atmosphere (less than a few days) and the other (of several years) when ocean and/or ice are involved. It sometimes exhibits a chaotic behavior, typical of nonlinear dynamical systems and difficult to predict deterministically. A major effort of fundamental research is still absolutely necessary.

2. A strong observational program, both regional and global

It is clear from our previous section that a strong observational program both at the regional and the global scales is absolutely indispensable to better climate understanding. An operational observing system should ideally:

1. seek integrated approaches that efficiently combine both remote and *in situ* observations;
2. integrate observational strategies in the terrestrial, oceanic, and atmospheric compartments, and appropriate paleo and human dimensions components;
3. be robust enough to provide long-term (years to decades) accurate, calibrated, and continuous measurements of all relevant parameters;
4. forge international collaboration so that harmonization of data, measurement and reporting procedure of the various national programs are well coordinated and the efforts of our distributed community are comparable and cumulative;
5. build an end-to-end data system that provides open access to the scientific community and the public, and rapidly evolves to exploit innovative technological developments; and
6. use these data sets in correcting and improving regional and global climate models, both in a diagnostic sense and in a predictive mode.

The list of required observations is necessarily long but can be grouped according to the research imperatives and the scientific questions emanating from them to advance our scientific understanding of the problem. For example, accurate long-term measurements of sea level alone cannot provide as much useful information as when combined with upper ocean temperatures, wind stress, sea ice extent, continental glacier changes, and upper ocean salinity. Similarly, cloud cover observations should be combined with water vapor and thermal profiles in the atmosphere, state of the land surface, and amounts of soil moisture. Another example is that of the atmospheric CO_2, whose spatial and temporal variations have to be studied in conjunction with land and ocean ecosystem types, surface temperatures, rates of land use change, and, of course, global circulation models.

The importance of accuracy, cross-calibration, and continuity over decades combined with technological innovations cannot be overstated. This is a tall order, and there is no single agency in the world that can carry out this agenda by itself. According to Wood and Fellous (Chapter 12), in Europe, the *Global Monitoring for Environment and Security* (GMES) initiative launched in 1998 by the European Commission (EC) and European Space Agency (ESA) has moved from concept to reality, with a series of major funding initiatives since 2001. The United States has also initiated a coordinated, national effort to achieve a US Integrated Earth Observation System (IEOS), which addresses several high-priority goals including the area of climate. It should be stressed that these efforts will get their full value only if the continuity of observations and measurements is ensured.

3. A serious assessment of sources of energy available today for global consumption

Some simple and well-accepted facts are that the Earth's population is increasing rapidly and will grow from 6.5 billion today to at least 7.5 billion in 2025 and likely up to 9 billion in 2050, based on models that use present cohort (i.e. age-aggregated) population distributions. It is also true that much of the population growth will occur in developing countries. These countries will therefore have to face a double challenge: ensuring their own economic development and providing enough food and energy to their growing populations.

The world energy needs will increase. Some scenarios of the International Energy Agency from OECD assume an energy demand growth rate of 1.6%–2% per year until 2030. This means that the energy demand will increase by about 50% by 2030 and be close to 16 000 Mtep (megatons of equivalent petroleum) to be compared to a value of about 10 500 Mtep in 2002. Much of this increase in demand will emerge from several rapidly developing nations: near 2030, the needs of China and India jointly will represent about 20% of the world's energy demand.

Resources of oil and gas are not infinite. Proven reserves of petroleum ensure their availability until about 2050. Scientific and technical progress will probably allow the discovery of new proven oil reserves, particularly if prices increase. Nevertheless, global reserves are dwindling, and increasing demand for finite petroleum resources cannot be a satisfactory policy for the long term.

Coal is abundant at the earth surface or at moderate depth under the ground. Proven resources ensure its availability for about two to three centuries. However, this energy source is far from being clean: To produce the same amount of energy, the emission of CO_2 from coal burning is about twice as much as from natural gas, the less carbon-intensive of the hydrocarbons. In addition, coal burning emits soot, dusts, and other pollutants, noticeably sulfur (responsible for significant urban pollution in winter) and heavy metals, which are highly toxic. Using extensively coal while maintaining friendly environmental conditions is difficult and will require new clean coal technologies to capture CO_2 emissions and eventually sequester the highly concentrated CO_2 thus captured.

Natural gas seems a contender as a bridge fuel to replace petroleum and coal in the near future until renewable alternatives have been found and developed. Some concerns remain, however, regarding its transport and distribution. Nevertheless, it is a significant candidate as a replacement fuel.

Nuclear power reactors produced about 6.5% of the world's energy needs in 2002, because nuclear power plants are used only to generate electricity, whereas the heat produced concurrently is lost. This energy source is poorly accepted by society in several countries. It requires a high technological level of knowledge and a continuous effort to ensure safety, with supervision by a strong, independent safety authority (this is the case in France, but was not in the former USSR). If nuclear energy were to develop with the present-day power plants (essentially Pressurized Water Reactors, PWR), it would consume a significant amount of the proven reserves of U^{235} within less than one century. A long-term policy using nuclear energy would require using breeder reactors, which essentially use either U^{238} as fuel, for which there is 140 times as much reserve as there is U^{235}. Another possibility would rest on Th^{232}, which is about three times more abundant than U^{238}, but this technology is still experimental. It should be pointed out, however, that nuclear power plants produce wastes, which are highly radioactive at first. The most radioactive isotopes decay within about 30 years, but long-life radioactive isotopes force humans to handle nuclear wastes for long-term storage and keep them isolated from the population for several hundred thousand years.

Renewable energy sources capture their energy from existing flows of energy, derived generally from solar energy through ongoing natural processes, such as solar radiation, wind, wave power, and flowing water (hydropower); from biological processes such as anaerobic digestion; and from geothermal heat flow. An inherent difficulty with renewable energies is their variable and diffuse nature (with the exception of geothermal energy). The geographic diversity of resources is also significant. Some nations have large resources, but at great distance from the major population centers, where the highest demands for electricity exist. One recurring criticism of renewable sources is their intermittent nature. Another one is the lack of scalability of existing systems to a level that would make a significant difference as replacement energy. Fossil fuels are easier to handle and have thus far been favored by industry. As a consequence, in 2002, renewable energy sources altogether accounted for only 12% of the world's energy needs (and this number includes noncommercial wood burning).

Given all of the above and the population increase, it is clear that the energy context will become increasingly tight during the next few decades. Therefore, the energy needs should be satisfied with well-known sources, already available at the industrial level, whereas oil and gas will become rarer.

4. A gigantic effort to develop an ambitious energy program for the next several decades

We are facing a major energy challenge for the next decades with contradictory requirements: providing more energy to a growing population and reducing both local and global emission of greenhouse gases. It should be pointed out that the Kyoto Protocol, which appears to be difficult to put in practice, does not allow stabilizing the atmospheric CO_2 concentration. The efforts, which are required to ensure a sustainable development during the twenty-first century, are much larger than the Kyoto requirements. Mid twenty-first-century primary energy requirements met by energies that are free of carbon dioxide emissions could be several times what we now derive from fossil fuels and will necessarily reside in a multifaceted approach of the new energy policy (see a detailed discussion in [148]).

A long-term strategy will therefore require:

1. Saving energy and developing new and efficient technologies.

Objectives are to live better with less energy consumption. This could begin with available technology if governments provide the necessary incentives, noticeably for urban heating and cooling, transportation, and low-energy equipments.

2. Using cleaner technologies for energy production.

Approximately one-third of all CO_2 emissions due to human activity come from fossil fuels used for generating electricity. CO_2 emissions will increase if coal replaces oil or gas. A variety of other industrial processes (oil refineries, cement works, and iron and steel production) also emit large amounts of CO_2 from each plant. All these emissions could be reduced substantially, without major changes to the basic process, by capturing and storing the CO_2 under the Earth's surface, for instance in old mines or natural geological oil reservoirs. It should be pointed out that this technology is just beginning at the experimental level (1 Mt of CO_2), whereas a significant reduction around the year 2050 would imply a sequestration of a quantity 10 000 times larger every year.

Nuclear energy has the potential to experience significant progress, but it will be small within the next few decades. Fusion aims at combining hydrogen atoms to get helium, as does the Sun to generate its energy. If we could use nuclear fusion reactions, we would have energy for millions of years. However, after 50 years of research, projects like ITER have only the goal to demonstrate from a scientific and technical point of view that fusion can potentially be used as an energy source on Earth. None of the projects is close to designing a plant, and it would be unrealistic to expect an industrial development of fusion within the next 50 years. Using more conventional fission reactions, Generation IV nuclear reactors would employ a closed fuel cycle to maximize the resource base and minimize high-level wastes to be stored in a repository. These reactors have still to be developed within international collaboration. The most optimistic view assumes that the beginning of development of this new technology will require at least 20 years of active research. We can thus count only on the present generation of nuclear reactors, which produce a significant amount of radioactive wastes but are free of CO_2 emission.

3. Reducing the emissions from transportation.

Transportation is the largest user of petroleum and, without drastically changing the mode of transportation of people and goods, there is no real alternative to it until hydrogen becomes a fuel. It is nevertheless the sector where the largest growth in energy usage and CO_2 emissions are predicted over the next 50 years, and these will come with mostly from developing countries, in particular China. It is important that fuel-efficient technology be put in place in both developed and developing countries. This will require a significant adjustment in some developed countries (the United States in particular) where automobile usage is extensive and the trend of using heavy vehicles has kept the fuel efficiency low despite technological improvements over the last decade or so.

Hydrogen might be used as a clean fuel for transportation within the next decades, replacing oil. However, it is presently derived from oil. Obviously, such a fuel should be produced by water electrolysis in large plants, which would consume enormous quantities of energy by themselves!

4. Developing actively renewable energy

Finally, the only infinite source of energy is the Sun. Solar energy in its various forms (direct, wind, photosynthesis, tides, etc.) should be the way of reference to prepare our adaptation to a world where fossil fuels will become scarce and expensive. A 'genesis' strategy requires developing all of these sources, noticeably directing utilization of sunshine, wind farms, or biofuels. This effort is just beginning. The US Department of Energy has set goals to replace 30% of the liquid petroleum transportation fuel with biofuels. A European Union Directive targets 5.75% of all petroleum and diesel transport fuels to be biomass-derived by 2010. These targets are achievable with present-day science and technology. Advances in genetics, biotechnology, process chemistry, and engineering might lead to a new manufacturing concept for converting renewable biomass to valuable fuels and products and have a positive impact on the economic and environmental well-being of society. However, care should be taken to avoid producing biofuels at the expense of food for the growing population.

Three major conclusions can be derived from this analysis: first, new technologies able to produce high amount of energy will not be available for at least 50 years, even if they are recognized as a top priority. Second, energy saving is absolutely necessary. Third, saving will not be sufficient, and a major technological effort is required to make the best use of known energy sources. Accordingly, we should not reject any source of energy, free of CO_2 emissions, for political or pseudoecological reasons.

5. Getting ready for the indispensable adaptation to future changes in the climate system

We hope that in the next few years the uncertainties will be reduced to such an extent that we would begin to place more confidence in the future scenarios of both regional and global climate models. But even before we get there we are sure of the following events:

1. As mentioned before, the world population would have increased by at least 1 billion people in the year 2030, reaching the level of more than 7 billion. More than half of this population will be living within 100 km from the coastline. Even if the intensity of extreme events remains as it is, battering of coastal zones by hurricanes and typhoons, the impending sea level rise, and overflow of the rivers in these regions need our full attention *now*!
2. Farming is another area that is very sensitive to droughts and floods on a regional scale and less sensitive to global climatic changes. It is therefore crucial that advanced regional weather forecasting systems be fully coordinated with agricultural networks, especially in emerging countries, to make the most efficient use of available water through storage and irrigation systems.
3. Freshwater availability to about a quarter of the world is currently in dire straits. This is mostly in Africa and South Asia when the summer monsoon is weak. With the dependence of these monsoons on El Niño and La Niña, it should now be relatively easy to forecast their intensity a year in advance. International aid organizations should try to assess how this newly acquired predictability can help to store rainwater in the years of heavy monsoons to serve the local population in the years of scarce rainfall.

As concluded by Balstad and Hourcade (Chapter 13), climate and climate variability have become an issue through which the plans and policies, the hopes and aspirations of individuals and nations are – and will be more and more – filtered.

It has the potential to alter 'reality' as we have known it in the past and to introduce changes not only in the physical world we inhabit, but also in our social and economic uses of that world. It links all the peoples of the Earth in a common system and forces us to consider responses to that system as a common endeavor. In this endeavor, French and American social scientists must work side by side with other scientists.

From this discussion, a message comes through 'loud and clear': adaptation of life in a changing climate is absolutely necessary. Scientists have first drawn attention to the public and policy makers on the problem of climate change. They were hardly believed initially, and it is only after the large impact of extreme events that some action was taken by policy makers (although there is no certainty that these events were the result of the anthropogenic climate change). Early in the 1990s, a program of "Industrial Ecology" was proposed by Socolow *et al.* [149] that was 'intended to mean both the interaction of global industrial civilization with the natural environment and the aggregate of opportunities for individual industries to transform their relationships with the natural environment.' It was, and still is, a great idea. This should be a central theme of our environmental policy in the twenty-first century and the backbone of our scientific 'genesis' strategy. It is certainly the cheapest and most cost-effective way to prepare for the future. Is it too much to wish that this message of the scientific community be seriously heard by decision makers before society faces monumental problems which will then be intractable under the pressure of events?

Glossary

°C. Temperature measured in degrees centigrade.

°F. Temperature measured in degrees Fahrenheit.

Abyssal flow. Refers to circulation in the deep ocean. It is related to a set of thermodynamic and dynamic processes called the thermohaline circulation.

Accretion. A process of (rain droplets) growth by accumulation and adherence.

Aerosols. Airborne (solid or liquid) particles from pollution, soil, vegetation, or marine origin.

Aerosol direct effect. Aerosol effect on the Earth radiation budget through scattering and absorption of solar radiation, generally resulting in a cooling of the Earth's system.

Aerosol indirect effect. Change in cloud microphysical and optical properties due to the role of aerosols as cloud condensation nuclei and ice nuclei, on which water droplets and ice crystals can form.

Aerosol semidirect effect. Absorption of solar radiation by aerosols, resulting in a change in the vertical temperature structure of the atmosphere with potential effects on cloud formation.

Albedo. The albedo of an Earth surface element is the fraction of solar energy reflected from the Earth back to space. It measures the reflectivity of the Earth's surface. Ice and snow have a high albedo and reflect up to 80% of solar energy, while forests and open ocean surface have a small albedo. The average Earth albedo is about 30%.

Altimetry (satellite altimetry). A satellite technique whereby the distance between the satellite and the sea surface is measured by a radar altimeter; this measure, combined with an accurate knowledge of the satellite orbit, provides the sea-surface elevation that is linked to ocean currents and many other ocean processes.

Arctic. Generally described as the region between the North Pole and the northern timberlines of North America and Eurasia. However, for scientific research, the region is often described in terms of the boundaries essential to the processes being studied.

Arctic Climate Impact Assessment (ACIA). The Arctic Climate Impact Assessment was established by the Arctic Council and the International Arctic Sciences Committee to evaluate and synthesize knowledge on climate variability and change and increased ultraviolet radiation, and support policy-making processes by addressing environmental, human health, social, cultural, and economic impacts and consequences, and to prepare policy recommendations. It was conducted by a team of over 300 scientists from 15 countries. The reports are available at www.acia.uaf.edu.

Arctic Council. The Arctic Council is an intergovernmental forum for addressing many of the common concerns and challenges faced by the Arctic states: Canada, Denmark (including Greenland and the Faroe Islands), Finland, Iceland, Norway, the Russian Federation, Sweden, and the United States. The Council is a unique forum for cooperation between eight national governments and these six indigenous peoples' organizations: Aleut International Association; Arctic Athabaskan Council; Gwich'in Council International; Inuit Circumpolar Conference; Russian Association of Indigenous Peoples of the North; Saami Council.

B2 and A2 emissions scenarios. Intergovernmental Panel on Climate Change (IPCC) developed a range of scenarios that plausibly describes future conditions of the planet Earth. The scenarios take into account future populations estimates, various projections of economic conditions, estimate of environmental conditions, issues of equity among societies and nations, technology projections, and projections on the stage of globalization. The B2 and A2 scenarios bracket the upper (A2) and the lower (B2) range of the middle third of all IPCC projected scenarios.

Carbon dioxide and other greenhouse gases. The major natural greenhouse gases are water vapor (H_2O), carbon dioxide (CO_2), methane (CH_4), ozone (O_3), nitrous oxide (N_2O), sulfur hexafluoride (SF_6), hydrofluorocarbons (HFC), perfluorocarbons (PFC), and chlorofluorocarbons (CFC).

Catchment. The region over which a stream collects its water (through runoff and tributaries).

CERES (Cloud and the Earth Radiant Energy System). Ensemble of instruments on board NASA satellites measuring the backscattered solar flux and the infrared radiation emitted by the Earth and its atmosphere.

Chlorofluorocarbons (CFCs). Industrially manufactured chemical compounds derived from the substitution of hydrogen by chlorine and fluorine atoms in hydrocarbons. These products were used in refrigeration systems and in many other domestic and industrial applications. Their production has been banned following the Montreal Protocol on the Protection of the Ozone Layer and its subsequent amendments.

Cirrus clouds. Typical layer-like ice cloud found in the upper troposphere.

Clausius-Clapeyron Law. Relationship between ambient temperature and the partial pressure of a particular species of volatile molecules (proportional to molecular number density) in the gaseous phase in equilibrium with the same molecular liquid. The equilibrium partial pressure depends on the chemical species and grows very quickly with increasing temperature.

Climate feedback. Positive or negative increment to the original forcing due to the adjustment of a particular component of the Earth environment to the change in global climatic parameters: surface and air temperature, humidity, winds, rainfall, etc.

Climate forcing. Incremental change in a term of the planetary radiation budget from some agreed-upon (climatological mean) reference value.

Climate sensitivity factor. Surface warming associated with a specified increment or 'forcing' of the energy budget of the planet (watts per square meter). The sensitivity factor very much depends on the range of further adjustments taken into account. Such adjustments are expected in the course of time, as the initial imbalance reaches other components of the Earth environment, each adding its own positive or negative contribution or 'feedback' to the original disturbance.

Cloud condensation nucleus (CCN). Aerosol particle on which water vapor can adhere and accumulate to form water or ice droplets from which clouds are made up.

Cloud microphysical processes. Range of physical-chemical processes that occur at the surface of preexisting liquid or solid particles, leading to the accretion or loss of water molecules, the growth or evaporation of raindrops, the formation or sublimation of ice particles, and, in general, the condensation of water vapor in either solid or liquid precipitation.

CNES. French space agency.

Convective clouds (cells). Relatively compact three-dimensional cellular circulation systems that develop in a conditionally unstable atmosphere and can shoot up to the top of the troposphere or beyond in the form of a characteristic billowing cloud tower. The life cycle of a large cell is on the order of half an hour and its horizontal extent on the order of 10 kilometers.

Cryosphere. Snow- and ice-covered areas. That part of the Earth's crust and atmosphere subject to temperatures below 0 °C (32 °F) for at least part of each year.

Cumulus clouds. Typical transient billowing clouds that may grow into major penetrating convective cloud towers spanning the whole depth of the troposphere.

Cyclonic Systems. See weather systems.

Deepwater formation. When the surface water gets denser than underlying waters, it sinks to depth. This change of density occurs when the water gets colder (e.g. in winter at high latitudes) or saltier (e.g. when sea ice is formed that leaves salt in the water). This deepwater formation is a local phenomenon, also described as deep convection.

Detrainment. Inverse process of air parcel entrainment by the turbulent eddies created by a strong jet – such as the ascending flow at the core of a convective cloud cell – penetrating a quiescent air mass.

Effective radius. A radius value used in place of the geometric radius to account for the distribution of size of particles (aerosols or cloud droplets).

ESA. European Space Agency.

Feedback. A feedback is a process whereby the effects of an original forcing feeds back into the original process, increasing or decreasing the final effects.

Flow and mixing in the ocean. The ocean currents can be described by separating what happens on relatively large distance scales (e.g. approximately above 50 km on the horizontal in the open ocean) and what happens on smaller scales. Mixing is used to refer to the smaller turbulent phenomena.

Geostationary orbit. Circular geocentric orbit at 35 786 km altitude above the Equator, in the equatorial plane, and with a zero eccentricity. A satellite in geostationary orbit has an orbital period exactly equal to the period of rotation of the Earth, i.e. 23 hours, 56 minutes, and 4 seconds, so that it appears fixed with respect to any point at the Earth surface.

Global Climate Models (or GCMs). A global climate model or general circulation model (GCM) generally combines numerical representations of the mechanical and physical laws that govern the atmospheric general circulation and the ocean circulation. Such models also include prescriptions as regards the evolution of the chemical, hydrological, or even biological characteristics of the atmosphere, oceans, and terrestrial surfaces, as well as glaciers and sea ice.

Global Warming Potential (GWP). The Global Warming Potential is the ratio of the global warming integrated over time expected from the injection of a given mass of a greenhouse gas to that expected from the same mass of carbon dioxide, taking into account the different absorption coefficients and residence time of the species. It is an estimate of the power of the gas to warm the planet compared to that of carbon dioxide. For example, over a 100-year span, methane has about 22 times the warming effect on the planet than does carbon dioxide.

Greenhouse effect. Absorption and re-emission of infrared radiation from the surface by the absorbing molecules or particulate matter in the atmosphere above. The greenhouse effect of absorbing gases or particles increases rapidly as the temperature difference between the absorbing medium and the surface increases.

Hadley circulation. Principal feature of the zonal mean circulation of the planetary atmosphere in a latitude-altitude cross-section. The Hadley circulation includes ascending motion at the Equator balanced by descending motions over the North and South tropical zones. It is characterized by a low-level inflow of air from the tropics toward the Equator and an outflow at high altitude in the upper troposphere.

Halons. Chemical compounds derived from the substitution of hydrogen by chlorine, bromine, and fluorine atoms in hydrocarbons; used, for example, in certain types of fire extinguishers.

IASC. International Arctic Sciences Committee.

Intergovernmental Panel on Climate Change (IPCC). The role of the IPCC is to assess on a comprehensive, objective, open, and transparent basis the scientific, technical, and socioeconomic information relevant to understanding the scientific basis of risk of human-induced climate change, its potential impacts, and options for adaptation and mitigation. IPCC reports should be neutral with respect to policy, although they may need to deal objectively with scientific, technical and socioeconomic factors relevant to the application of particular policies.

–From the IPCC Terms of Reference of the United Nations Framework Convention on Climate Change

International Arctic Sciences Committee (IASC). The International Arctic Science Committee is a nongovernmental organization whose aim is to encourage and facilitate cooperation in all aspects of Arctic research, in all countries engaged in Arctic research, and in all areas of the Arctic region. Its members are appointed by the academies of science from 19 nations.

IPSL. Institut Pierre-Simon Laplace.

ITCZ. Inter-Tropical Convergence Zone.

Kyoto Protocol. The Kyoto Protocol to the United Nations Framework Convention on Climate Change is an amendment to the international treaty on climate change, assigning mandatory targets for the reduction of greenhouse gas emissions to signatory nations.

Lidar. Instrument using short light pulses emitted by a laser source to measure a distance or probe the atmosphere (or both).

Meridional. In the north-south direction.

MERIS. The Medium Resolution Imaging Spectrometer is a programmable, medium-spectral resolution imaging spectrometer measuring the solar radiation reflected by the Earth. Fifteen spectral bands can be selected by ground command, each of which has a programmable width and a programmable location in the 390–1040 nm spectral range.

Meso-scale. Intermediate horizontal scales of phenomena ranging from individual convective cloud (10 km) to major weather compact systems such as tropical cyclones (several hundred kilometers).

Mesosphere. Region of the atmosphere located from approximately 50 km to 85–90 km altitude, characterized by a decrease of temperature with height. The top of the mesosphere is called the mesopause, where temperature reaches a minimum of approximately −150 °C (−235 °F) at the summer pole.

MISR. The Multiangle Imaging SpectroRadiometer is an instrument imaging Earth's system simultaneously at nine different angles in the visible spectral region.

Microwave radiometer. A sensitive electronic instrument that measures the amplitude (or strength) of microwave signals at GHz frequencies (i.e. at mm wavelengths).

MODIS. The Moderate resolution Imaging System is an instrument imaging Earth's atmosphere and surface at about 30 wavelengths ranging from the near ultraviolet to the thermal infrared emitted radiation.

Moorings. One of the techniques to observe the ocean with time. The instruments are placed on a line that is anchored at the bottom and extends toward the surface. Typical measuring devices are current meters that measure the ocean current velocity and direction.

NASA. US National Aeronautics and Space Administration.

Navier-Stokes equation. Equation that describes the balance of forces and acceleration applied to a fluid parcel (equivalent to the Newton equation for a punctual mass). It is the fundamental equation for ocean and atmosphere dynamics.

NOAA. US National Ocean and Atmosphere Administration.

Northern Sea Route. The Northern Sea Route is a shipping lane from the Atlantic Ocean to the Pacific Ocean along the Russian coasts of the Far East and Siberia.

The vast majority of the route lies in Arctic waters. Before the beginning of the twentieth century, it was known as the Northeast Passage.

Nutrients. Describes a set of chemical species that are fundamental sources of nutrition to the ecosystems. In the ocean, major nutrients are nitrates, phosphates, and dissolved silica, whereas minor nutrients like iron are also necessary.

OMI. The Ozone Monitoring Instrument measures the backscatter radiation in the visible and ultraviolet from which ozone and other air quality components such as NO_2, SO_2, BrO, OClO, and aerosol characteristics can be estimated.

Optical sensor. A sensitive electronic instrument that measures the amplitude (or strength) of optical signals; that is, at visible wavelengths.

Optical thickness. Cumulative effect of the absorption of electromagnetic radiation through an absorbing medium. The radiation attenuation follows an exponential law, with the optical depth as the exponent.

Orography. The nature of a region with respect to its elevated terrain.

OSTM. Ocean Surface Topography Mission (the NASA-CNES Jason-2 mission, to be launched in 2008).

Outlet glaciers. An outlet glacier is a valley glacier that drains the ice sheet near the coast. In Antarctica, 80% of the ice is drained by a few tens of large outlet glaciers.

Permafrost. Permafrost (or permafrost soil) is a condition in which ground material remains frozen – i.e. stays at or below 0 °C (32 °F) for two or more years. Most permafrost is located in high latitudes (e.g. North and South Poles), but alpine permafrost exists at high altitudes. The extent of permafrost can vary as the climate changes. Today, approximately 20% of the Earth's land mass is covered by permafrost (including discontinuous permafrost) or glacial ice. Overlying permafrost is a layer of ground called the active layer that seasonally thaws during the summer. Plant life can be supported only within the active layer, because growth can occur only in soil that is fully thawed for some part of the year. Thickness of the active layer varies by year and location but is typically 0.6–4.0 m (2–12 feet) thick. In areas of continuous permafrost and harsh winters, the depth of the permafrost can be very great: 440 m (1330 feet) at Barrow, Alaska; 600 m (1970 feet) at Prudhoe Bay, Alaska; up to 726 m (2382 feet) in the Canadian Arctic islands; and as much as 1493 m (4510 feet) in the northern Lena and Yana River basins in Siberia.

Phenology. Phenomena concerned with the temporal development, particularly the annual cycle, of fauna and flora.

Planetary boundary layer. Atmospheric layer where aerodynamic turbulence created by friction over ocean or ground surfaces has a controlling effect on vertical motions and mixing. The planetary boundary layer extends up to a 1- to 2-km altitude, above which the 'free atmosphere' circulation is largely immune to the effects of boundary turbulence.

Polder (Polarization and Directionality of the Earth's Reflectance). A wide field of view imaging radiometer that measures backscattered radiation at different visible wavelengths and for different polarizations, from which cloud, aerosols, and surface properties can be determined.

Poleward. Toward the pole.

Ppm(v), ppb(v). Part per million, part per billion (in volume).

Precipitable water. Total mass of water vapor per surface unit present in a vertical air column spanning the whole depth of the atmosphere. Global-mean total precipitable water in the atmosphere is equivalent to a 2.5-cm or 1-inch-thick layer of condensed water.

Radar. Active instrument emitting radioelectric impulsions and analyzing received echoes from a target to derive information on its distance and surface characteristics.

Radiative forcing. Change in the Earth's radiative budget due to an external climate agent, such as anthropogenic greenhouse gases or aerosols, since preindustrial times. The radiative forcing can be considered as an incremental energy input to the system composed by the troposphere, oceans, the cryosphere, and the continents. This input must be balanced by the excess energy storage (for example, in the ocean) or by an increase in the outgoing infrared radiation from the tropopause.

Radiometer. Sensor that quantitatively measures the intensity of electromagnetic radiation in a particular wavelength or band of the electromagnetic spectrum. For example, remote sensing satellites carry microwave radiometers, used in meteorological or hydrological applications, or to measure sea ice extent, optical multispectral radiometers to observe vegetation cover, or infrared radiometers to measure surface temperature.

Relative humidity. Ratio of the number density of water molecules actually present in the air to the maximum molecule density that could subsist in gaseous state at that particular temperature.

Scavenging. Process by which soluble chemical compounds are removed from the atmosphere through precipitation.

SMOS (Soil Moisture and Ocean Salinity). An ESA satellite mission to be launched in 2008.

SST. Sea surface temperature.

Stratopause. Upper limit of the stratosphere situated around 50 km and corresponding to a local maximum of temperature of about 0 °C (32 °F).

Stratosphere. Region of the atmosphere situated between the tropopause and the altitude of approximately 50 km. This region of the atmosphere in which the ozone layer is located is characterized by a gradual increase of temperature with height ensuring a high stability of air masses vis-à-vis vertical motions.

Stratus clouds. Typical low-altitude liquid water cloud, characterized by a layered structure with sharp vertical boundaries and vast horizontal extent.

Subsidence. Descending atmospheric motions.

Sun-synchronous orbit. Circular geocentric orbit with an altitude and inclination chosen in such a way that the angle between the orbit plane and the direction of the Sun remains nearly constant, in spite of the annual drift (precession) of the orbit plane. A satellite in sun-synchronous orbit overpasses a given location on the Earth's surface at the same local solar time, so that the solar zenith angle (i.e. the scene illumination) varies little from one pass to the next. Sun-synchronous orbits are possible in an altitude range of 600–1000 km, with orbit revolution periods of 96–100 minutes and inclination of about 98° to the Equator.

Synthetic Aperture Radar (SAR). An active microwave sensor on board a mobile platform (aircraft, satellite) providing high resolution images of the Earth surface. It works by collecting the echo returns from many radar pulses and processing them into a single radar image.

Thermodynamics. Names of the processes that describe exchange and transformation of thermal energy. A key thermodynamic process is the exchange of heat and freshwater between the ocean and the atmosphere.

Tide gauges. Instruments that measure the sea level change, either by measuring the sea surface elevation or by measuring underwater pressure. Primarily designed to monitor sea level changes related to tides, they also detect variations of ocean currents, ocean heat content, and ocean water mass.

TOGA. Tropical Ocean and Global Atmosphere, an experiment of the World Climate Research Programme (1985–1995).

TOMS. The Total Ozone Mapping System is a multichannel radiometer that measures backscattered ultraviolet radiation at different wavelengths, from which the total atmospheric ozone content can be determined.

Tropopause. Boundary layer between the troposphere and the stratosphere situated near 15–18 km (45 000–54 000 ft) in the equatorial zone, where the temperature is of the order of −70 to −80 °C (−92 to −110 °F). In the extra-tropics, the altitude of the tropopause is lower (typically 8–12 km or 24 000–6 000 ft), and its mean temperature is somewhat higher than in the tropics.

Troposphere. Region of the atmosphere situated between the surface and to an altitude of 8–18 km (24 000–54 000 ft), depending on the latitude. It is characterized by a rapid decrease of temperature as a function of height (−6.5 °C/km or −11.6 °F/km), a relatively low stability that facilitates vertical motions, and a strong mechanical and thermodynamic activity. The troposphere is the motor of all other climate system components.

United Nations Environment Program (UNEP). The United Nations Environment Program was established in 1972 by the UN to provide leadership and encourage partnership in caring for the environment by inspiring, informing, and enabling nations and peoples to improve their quality of life without compromising that of future generations.

United Nations Framework Convention on Climate Change (UNFCCC). The United Nations Framework Convention on Climate Change (UNFCCC or FCCC) is an international environmental treaty produced at the United Nations Conference on Environment and Development, held in Rio de Janeiro in 1992. The treaty is designed to provide an international venue to establish mechanisms to reduce emissions of greenhouse gas in order to combat global warming.

Weather systems. Nonlinear disturbances that appear and grow spontaneously in the general circulation of the atmosphere until exhaustion of the available energy. These perturbations generate strong winds, major vertical motions up and down, and most rainfall and snowfall.

WMO-UNEP Scientific Assessment of Ozone Depletion. International assessment of the state of the stratospheric ozone layer, which takes place every 4–5 years under the auspices of the World Meteorological Organization and the United Nations Environmental Programme.

WOCE. World Ocean Circulation Experiment, an experiment of the World Climate Research Programme (1990–2000).

References

[1] J. T. Houghton, G. J. Jenkins and J. J. Ephraums, eds., *Climate Change: The IPCC Scientific Assessment* (Cambridge: Cambridge University Press, 1990). (The IPCC First Assessment Report, or FAR).

[2] J. T. Houghton, L. G. Meira Filho, B. A. Callander, N. Harris, A. Kattenberg, and K. Maskell, *Climate Change 1995. The Science of Climate Change* (Cambridge: Cambridge University Press, 1996). (The IPCC Second Assessment Report, or SAR).

[3] J. T. Houghton, Y. Ding, D. J. Griggs, et al., eds., *Climate Change 2001: The Scientific Basis* (Cambridge: Cambridge University Press, 2001a). (The IPCC Third Assessment Report, or TAR). All IPCC pdf files are available at: www.ipcc.ch/.

[4] R. T. Watson, ed., *Climate Change 2001: Synthesis Report. A contribution of Working Groups I, II, and III to the Third Assessment Report of the Intergovernmental Panel on Climate Change* (Cambridge: Cambridge University Press, 2001b). (The TAR Synthesis Report, including the Summary for Policymakers and the Working Group Summaries).

[5] Intergovernmental Panel on Climate Change, *Climate Change 2007: The Physical Science Basis. Contributions of Working Group I to the Fourth Assessment Report of the Intergovernmental Panel on Climate Change* (United Kingdom/New York: Cambridge University Press, 2007).

[6] S. Arrhenius, On the influence of carbonic acid in the air upon the temperature of the ground. *The London, Edimburgh and Dublin Philosophical Magazine and Journal of Science*, **41** (1896), 237–76.

[7] J.-B. J. Fourier, Remarques générales sur les températures du globe terrestre et des espaces planétaires. *Annales de Chimie et Physique, 2ème série*, **XXVI** (1824), 136–167.

[8] J. London, *A Study of the Atmospheric Heat Balance (Final Report – Research Division, College of Engineering, Dept. of Meteorology and Oceanography, New York University; Contract no. AF 19)* (New York: New York University, 1957).

[9] T. H. Vonder Haar and V. E. Suomi, Measurements of the Earth's radiation budget from satellites during a five-year period. Part 1: Extended time and space means. *Journal of Atmospheric Sciences*, **28** (1971), 305–14.

[10] B. R. Barkstrom, E. F. Harrison, G. L. Smith, R. N. Green, J. Kibler, R. Cess and the ERBE Science Team, Earth Radiation Budget Experiment (ERBE): Archival of April 1985 Results. *Bulletin of the American Meteorological Society*, **70**:10 (1989), 254–1262.

[11] R. Kandel, M. Viollier, P. Raberanto, et al., The ScaRaB Earth Radiation Budget Dataset. *Bulletin of the American Meteorological Society*, **79**:5 (1998), 765–83.

[12] V. Ramanathan, R. D. Cess, E. F. Harrison, P. Minnis, B. R. Barkstrom, E. Ahmad and D. Hartman, Cloud-radiative forcing and climate: insights from the Earth Radiation Budget Experiment. *Science*, **243** (1989), 57–63.

[13] B. J. Soden, Variations in the tropical greenhouse effect during El Niño. *Journal of Climate*, **10**:5 (1997), 1050–5.

[14] F. J. Doblas-Reyes R. Hagedorn and T. N. Palmer, The rationale behind the success of multi-model ensembles in seasonal forecasting. Part II: Calibration and combination. *Tellus A*, **57** (2005), 234–52.

[15] J. R. Christy, R. W. Spencer and W. D. Braswell, MSU tropospheric temperatures: Dataset construction and radiosonde comparisons. *Journal of Atmospheric and Oceanic Techniques*, **17** (2000), 1153–70.

[16] D. Randall, M. Khairoutdinov, A. Arakawa and W. Grabowski, Breaking the cloud parameterization deadlock. *BAMS*, **84** (2003), 1547–64.

[17] M. Wild, H. Gilgen, A. Roesch, et al., From dimming to brightening: decadal changes in solar radiation at Earth's surface. *Science*, **308** (2005), 847–50.

[18] B. A. Wielicki, T. Wong, N. Loeb, P. Minnis, K. Priestley and R. S. Kandel, Changes in Earth's albedo measured by satellite. *Science*, **308** (2005), 825.

[19] E. P. Pallé, R. Goode, P. Montañés-Rodríguez and S. E. Koonin, Changes in Earth's reflectance over the past two decades. *Science*, **304** (2004), 1299–1301.

[20] R. S. Lindzen, Can increasing carbon dioxide cause climate change? *Proceedings of the National Academy of Sciences of the USA*, **94** (1977), 8335–342.

[21] S. Manabe and R. T. Wetherald, Thermal equilibrium of the atmosphere with a given distribution of relative humidity. *Journal of Atmospheric Sciences*, **24** (1967), 241–59.

[22] M. Miller, A. Beljaars and T. Palmer, The sensitivity of the ECMWF Model to the parameterization of evaporation from the tropical oceans. *Journal of Climate*, **5** (1992), 418–34.

[23] T. J. Phillips, G. L. Potter, D. L. Williamson, et al., Evaluating parameterizations in general circulation models: climate simulation meets weather prediction. *Bulletin of the American Meteorological Society*, **85** (2004), 1903–15.

[24] K. E. Trenberth, J. Fasullo and L. Smith, Trends and variability in column-integrated atmospheric water vapor, *Climate Dynamics*, **24** (2005), 742–58.

[25] S. L. Dingman, *Physical Hydrology*, 2nd edn (Upper Saddle River, NJ: Prentice Hall, 2002).

[26] R. D. Brown, Northern Hemisphere snow cover variability and change: 1915–1997. *Journal of Climate*, **13**:13 (2000), 2339–55.

[27] D. A. Stow, A. Hope, D. McGuire, et al., Remote sensing of vegetation and land-cover change in Arctic Tundra Ecosystems. *Remote Sensing of Environment*, **89**:3 (204), 281–308.

[28] T. P. Barnett, J. C. Adam and D. P. Lettenmaier, Potential impacts of a warming climate on water availability in snow-dominated regions. *Nature*, **438** (2005), 303–9.

[29] R. J. Ross and W. P. Elliott, Radiosonde-based Northern Hemisphere tropospheric water vapor trends. *Journal of Climate*, **14** (2001), 1602–11.

[30] P. J. Webster, G. J. Holland, A. Curry and H. R. Chang, Changes in tropical cyclones number, duration and intensity in a warming environment. *Science*, **309** (2005), 1844–6.

[31] P. C. D. Milly and K. A. Dunne, Trends in evaporation and surface cooling in the Mississippi River basin. *Geophysical Research Letters*, **28** (2001), 1219–22.

[32] M. Biasutti and A. Giannini, Robust Sahel drying in response to late 20th century forcings. *Geophysical Research Letters*, **33** (2006), L11706.

[33] I. M. Held, T. L. Delworth, J. Lu, K. L. Findell and T. R. Knutson, Simulation of Sahel drought in the 20th and 21st centuries. Proceedings of the National Academy of Sciences, **102** (2005), 17,891–6.

[34] T. G. Huntington, Evidence for intensification of the global water cycle: Review and synthesis. *Journal of Hydrology*, **319** (2006), 83–95.

[35] N. Nakicenovic and R. Swart, eds., *Special Report on Emissions Scenarios*, Intergovernmental Panel on Climate Change (Port Chester, NY: Cambridge University Press, 2005). (see also www.grida.no/climate/ipcc/emission/index.htm)

[36] J. L. Redelsperger, C. D. Thorncroft, A. Diedhiou, T. Lebel, D. J. Parker and J. Polcher, African Monsoon Multidisciplinary Analysis (AMMA): An International Research Project and Field Campaign. *Bulletin of the American Meteorological Society*, 2006.

[37] R. A. Pielke, G. Marland, R. A. Betts, *et al.*, The influence of land-use change and landscape dynamics on the climate system: relevance to climate-change policy beyond the radiative effect of greenhouse gases. *Philosophical Transactions of the Royal Society of London, Serie A – Mathematical, Physical, and Engineering Sciences*, **360** (2002), 1797, 1705–19.

[38] H. Stommel and A. B. Arons, On the abyssal circulation of the world ocean II. An idealized model of the circulation pattern and amplitude in oceanic basins. *Deep-Sea Research*, **6** (1960), 217–33.

[39] A. Ganachaud, Large scale oceanic circulation and fluxes of freshwater, heat, nutrients and oxygen. Unpublished Ph.D. thesis, MIT/WHOI (1999).

[40] J. D. Barrow and S. P. Bhavasar, Filaments: what the astronomer's eye tells the astronomer's brain. *Quarterly Journal of the Royal Astronomical Society*, **28** (1987), 109–28.

[41] H. F. Diaz and V. Markgraf, *El Niño: Historical and Paleoclimatic Aspects of the Southern Oscillation* (Cambridge: Cambridge University Press, 1992).

[42] W. B. Broecker, The biggest chill. *Natural History*, **96** (1987), 74–82.

[43] J. Church, J. M. Gregory, P. Huybrechts, *et al.*, 2001. Changes in sea level. In [3]: J. T. Houghton, Y. Ding, D. J. Griggs, *et al.*, eds., *Climate Change 2001: The Scientific Basis* (Cambridge: Cambridge University Press, 2001a).

[44] B. C. Douglas, M. S. Kearney, S. P. Leatherman, eds, *Sea Level Rise – History and Consequences* (San Diego, CA: Academic Press, 2001).

[45] A. Cazenave and R. S. Nerem, Present-day sea level change: observations and causes. *Reviews of Geophysics*, **42** (2004), RG3001, doi:10–1029/04/2003RG000139.

[46] G. Holloway and T. Sou, Has Arctic sea ice rapidly thinned? *Journal of Climate*, **15** (2002), 1691–1701.

[47] D. A. Rothrock, J. Zhang and Y. Yu, The Arctic ice thickness anomaly of the 1990s: A consistent view from observations and models. *Journal of Geophysical Research*, **108**:C3 (2003), 3083–92.

[48] W. Meier J. Stroeve, F. Fetterer and K. Knowles, Reductions in Arctic sea ice cover no longer limited to summer. *EOS*, **86**:36 (2005), 326–7.

[49] J. T. Overpeck, *et al.*, Arctic system on trajectory to new, seasonally ice-free state, *EOS Transactions*, **86** (2005), 309–13.

[50] D. T. Shindell and G. A. Schmidt, Southern Hemisphere climate response to ozone changes and greenhouse gas increases. *Geophysical Research Letters*, **31** (2004), 1029.

[51] M. M. Holland and C. M. Bitz, Polar amplification of climate change in coupled models. *Climate Dynamics*, **21** (2003), 221–32.

[52] W. S. Broecker, Thermohaline circulation, the Achilles heel of our climate system: Will man-made CO_2 upset the current balance? *Science*, **278** (1997), 1582–8.

[53] Antarctic Climate Impact Assessment (ACIA), *Impacts of a Warming Arctic* (New York: Cambridge University Press, 2004). (available at http://www.amap.no)

[54] See http://nsidc.org/sotc/sea_ice.html.

[55] P. J. Crutzen and E. F. Stoermer, *IGBP Newsletter*, **41** (2000).

[56] P. J. Crutzen, Geology of mankind. *Nature*, **415** (2002), 23.

[57] J. M. Barnola, M. Anklin, J. Porcheron, D. Raynaud, J. Schwander and B. Stauffer, CO_2 evolution during the last millennium as recorded by Antarctic and Greenland ice. *Tellus*, **47** (1995), 264–72.

[58] D. M. Etheridge, L. P. Steele, R. L. Langenfelds, R. J. Francey, J.-M. Barnola and V. I. Morgan, Natural and anthropogenic changes in atmospheric CO_2 over the last 1000 years from air in Antarctic ice and firn. *Journal of Geophysical Research – Atmospheres*, **101** (1996), 4115–28.

[59] D. M. Etheridge, L. P. Steele, R. J. Francey and R. L. Langenfelds, Atmospheric methane between 1000 AD and present: Evidence of anthropogenic emissions and climatic variability. *Journal of Geophysical Research-Atmospheres*, **103** (1998), 15979–93.

[60] D. F. Ferretti, J. B. Miller, J. W. C. White, *et al.*, Unexpected changes to the global methane budget over the last 2,000 years. *Science*, **309** (2005), 1714–7.

[61] J.-M. Barnola, D. Raynaud, Y. S. Korotkevich and C. Lorius, Vostok ice core provides 160,000-year record of atmospheric CO_2. *Nature*, **329** (1987), 408–14.

[62] T. Blunier, J. A. Chappellaz, J. Schwander, B. Stauffer and D. Raynaud, Variations in atmospheric methane concentration during the Holocene epoch. *Nature*, **374** (1995), 46–9.

[63] D. Raynaud, J. Jouzel, J.-M. Barnola, J. Chappellaz, R. J. Delmas and C. Lorius, The ice record of greenhouse gases. *Science*, **259** (1993), 926–34.

[64] C. D. Keeling, S. C. Piper and M. Heimann, A three-dimensional model of atmospheric CO_2 transport based on observed winds: 4. Mean annual gradients and interannual variations, in aspects of climate variability in the Pacific and Western Americas. In *Geophysical monograph 55*, ed. D. H. Peterson (Washington DC: American Geophysical Union, 1989).

[65] R. Keeling, S. C. Piper and M. Heimann, Global and hemispheric CO_2 sinks deduced from changes in atmospheric O_2 concentration. *Nature*, **381** (1996), 218–21.

[66] A. C. Manning and R. F. Keeling, Global oceanic and land biotic carbon sinks from the Scripps atmospheric oxygen flask sampling network. *Tellus*, **58B:n°2** (2006), 95–116.

[67] P. Bousquet, P. Peylin, P. Ciais, C. Le Quéré, P. Friedlingstein and P. P. Tans, Regional changes in carbon dioxide fluxes of land and oceans since 1980. *Science*, **290** (2000), 1342–6.

[68] W. Lucht, I. C. Prentice, R. B. Myneni, *et al.*, Climatic control of the high-latitude vegetation greening trend and Pinatubo effect. *Science*, **296** (2002), 1687–9.

[69] C. Rodenbeck, S. Houweling, M. Gloor and M. Heimann, Time-dependent atmospheric CO_2 inversions based on interannually varying tracer transport. *Tellus Series B-Chemical and Physical Meteorology*, **55** (2003), 488–97.

[70] H. Q. Tian, J. M. Melillo, D. W. Kicklighter, *et al.*, Effect of interannual climate variability on carbon storage in Amazonian ecosystems. *Nature*, **396** (1998), 664–7.

[71] P. Ciais, P. P. Tans, M. Trolier, J. W. White and R. Francey, A large Northern Hemisphere terrestrial CO_2 sink indicated by the $^{13}C/^{12}C$ ratio of atmospheric CO_2. *Science*, **269** (1995), 1098–1102.

[72] S. Fan, Modelling long-range transport of CFCs to Mace Head, Ireland. *Quarterly Journal of the Meteorological Society*, **124** (1998), 417–46.

[73] K. R. Gurney, R. M. Law, A. S. Denning, *et al.*, Towards robust regional estimates of CO_2 sources and sinks using atmospheric transport models. *Nature*, **415** (2002), 626–30.

[74] P. J. Rayner and R. M. Law, The interannual variability of the global carbon cycle . *Tellus*, **51** (1999), 210–12.

[75] P. P. Tans, I. Y. Fung, and T. Takahashi, Observational constraints on the global atmospheric CO_2 budget. *Science*, **247** (1990), 1431–8.

[76] W. Peters, E. J. Dlugokencky, F. J. Dentener, *et al.*, Toward regional-scale modeling using the two-way nested global model TM5: Characterization of transport using SF_6. *Journal of Geophysical Research-Atmospheres*, **109** (2004), 19314.

[77] P. Peylin, P. J. Rayner, P. Bousquet, C. Carouge, F. Hourdin, P. Heinrich, P. Ciais, and Aerocarb contributors, Daily CO_2 flux estimates over Europe from continuous atmospheric measurements: 1. Inverse methodology. *Atmospheric Chemistry and Physics Discussions*, **5** (2005), 1647–78.

[78] P. Rayner, M. Scholze, W. Knorr, T. Kaminski, R. Giering and H. Widmann, Two decades of terrestrial carbon fluxes from a Carbon Cycle Data Assimilation System (CCDAS). *Global Biogeochemical Cycles*, **19** (2005), doi:10.1029/2004GB002254.

[79] J. C. Orr, E. Maier-Reimer, U. Mikolajewicz, *et al.*, Estimates of anthropogenic carbon uptake from four 3-D global ocean models. *Global Biogeochemical Cycles*, **15** (2001), n°11, 43–60.

[80] J. L. Sarmiento and E. T. Sundquist, Revised budget for the oceanic uptake of anthropogenic carbon dioxide. *Nature*, **356** (1992), 589–93.

[81] W. S. Broecker and T. H. Peng, *Tracers in the Sea* (Palisades, New York: Lamont-Doherty Geological Observatory, 1982).

[82] J. Orr, Accord of ocean models in predicting uptake of anthropogenic CO_2. *Water, Air & Soil Pollution*, **70** (1993), 465–81.

[83] L. Bopp, O. Aumont, S. Belviso and P. Monfray, Potential impact of climate change on marine dimethyl sulfide emissions, *Tellus*, **55B** (2003), 11–22.

[84] F. Joos, G.-K. Plattner, T. F. Stocker, O. Marchal and A. Schmittner, Global warming and marine carbon cycle feedbacks on future atmospheric CO_2. *Science*, **284** (1999), 464–7.

[85] D. E. Archer, H. Kheshgi and E. Maier-Reimer, Multiple timescales for neutralization of fossil fuel CO_2. *Geophysical Research Letters*, **24** (1997), 405–8.

[86] A. Indermühle, T. F. Stocker, F. Joss, *et al.*, Holocene carbon-cycle dynamics based on CO_2 trapped in ice at Taylor Dome, Antarctica. *Nature*, **398** (1999), 121–6.

[87] M. L. Goulden, J. W. Munger, S. M. Fan, B. C. Daube and S. C. Wofsy, Exchange of carbon dioxide by a deciduous forest: Response to interannual climate variability, *Science*, **271** (1996), 1576–8.

[88] S. C. Wofsy, M. L. Goulden, J. W. Munger, *et al.*, Net exchange of CO_2 in a midlatitude forest, *Science*, **260** (1993), 1314–7.

[89] P. Ciais, M. Reichstein, N. Viovy, *et al.*, An unprecedented reduction in the primary productivity of Europe during 2003 caused by heat and drought. *Nature*, **437** (2005), 529–33.

[90] R. A. Houghton, Revised estimates of the annual net flux of carbon to the atmosphere from changes in land use and land management 1850–2000, *Tellus B* (2003), 378–90.

[91] N. Ramankutty and J. A. Foley, Estimating historical changes in global land cover: croplands from 1700 to 1992. *Global Biogeochemical Cycles*, **13** (1999), 997–1028.

[92] K. Klein Goldewijk, Estimating global land use change over the past 300 years: the HYDE database. *Global Biogeochemical Cycles*, **15** (2001), 417–34.

[93] R. S. DeFries, R. A. Houghton, M. C. Hansen, C. B. Field, D. Skole and J. Townshend, Carbon emissions from tropical deforestation and regrowth based on satellite observations for the 1980s and 1990s. *Proceedings of the National Academy of Sciences*, **99**:n°22 (2002), 14256–61.

[94] F. Achard, H. D. Eva, P. Mayaux, H.-J. Stibig and A. Belward, Improved estimates of net carbon emissions from land cover change in the tropics for the 1990s. *Global Biogeochemical Cycles*, **18**, (2004), GB2008, doi:10.1029/2003GB002142.

[95] G. C. Hurtt, S. Frolking, M. G. Fearon, *et al.*, The underpinnings of land-use history: three centuries of global gridded land-use transitions, wood-harvest activity, and resulting secondary lands. *Global Change Biology*, **12**, (2006), 1–22.

[96] J. P. Caspersen, S. W. Pacala, J. C. Jenkins, G. C. Hurtt, P. R. Moorcroft and R. A. Birdsey, Contributions of land-use history to carbon accumulation in U.S. forests. *Science*, **290**: n°5494 (2000), 1148–51.

[97] S. W. Pacala, G. C. Hurtt, D. Baker, *et al.*, Consistent land- and atmosphere-based U.S. carbon sink estimates. *Science*, **292**:n°5525 (2001), 2316–20.

[98] S. J. Pyne, *America's Fires: Management of Wildlands and Forests* (Durham, NH: Forest History Society, 1997).

[99] E. Dwyer, S. Pinnock, J.-M. Grégoire and J. M. C. Pereira, Global spatial and temporal distribution of vegetation fire as determined from satellite observations. *International Journal of Remote Sensing*, **21** (2000), 1289–1302.

[100] Y. Malhi and J. Grace, Tropical forests and atmospheric carbon dioxide. *Trends in Ecology and Evolution*, **15** (2000), 332–7.

[101] D. Skole and C. J. Tucker, Tropical deforestation and habitat fragmentation in the Amazon: satellite data from 1978 to 1988. *Science*, **206** (1993), 1905–9.

[102] F. H. Achard, D. Eva, H.-J. Stibig, *et al.*, Determination of deforestation rates of the world's humid tropical forests. *Science*, **297**: n°5583 (2002), 999–1002.

[103] L. M. Curran, S. N. Trigg, A. K. McDonald, *et al.*, Lowland forest loss in protected areas of Indonesian Borneo. *Science*, **303** (2004), 1000–3.

[104] S. E. Page, F. Siegert, J. O. Rieley, H. D. V. Boehm, A. Jaya and S. Limin, The amount of carbon released from peat and forest fires in Indonesia during 1997, *Nature*, **420** (2002), 61–5.

[105] B. J. Stocks, J. A. Mason, J. B. Todd, *et al.*, Large forest fires in Canada, 1959–1997. *Journal of Geophysical Research (Atmospheres)*, **108**:D1 (2002), 8149.

[106] G. R. Van der Werf, J. T. Randerson, G. J. Collatz, *et al.*, Continental-scale partitioning of fire emissions during the 1997 to 2001 El Niño /La Niña period, *Science*, **303** (2004), 73–6.

[107] D. Mollicone, H. D. Eva and F. Achard, Human role in Russian wildfires. *Nature*, **440** (2006), 436–7.

[108] G. C. Hurtt, S. W. Pacala, P. R. Moorcroft, *et al.*, Projecting the future of the U.S. carbon sink. *Proceedings of the National Academy of Sciences of the USA*, **99**:n°3 (2002), 1289–1394.

[109] K. A. Kvenvolden, Potential effects of gas hydrates on human welfare. *Proceedings of the National Academy of Sciences of the USA*, **96** (1999), n°7, 3420–6.

[110] A. V. Milkov, Global estimates of hydrate-bound gas in marine sediments: how much is really out there? *Earth Science Reviews*, **66** (2004), 183–97.

[111] R. B. Myneni, C. D. Keeling, C. J. Tucker, G. Asrar and R. R. Nemani, Increased plant growth in the northern high latitudes from 1981 to 1991. *Nature*, **386** (1997), 698–702.

[112] D. W. Kicklighter, M. Bruno, S. Donges, *et al.*, A first order analysis of the potential role of CO_2 fertilization to affect the global carbon budget: a comparison of four terrestrial biosphere models. *Tellus*, **51B** (1999) 343–66.

[113] Prentice, I. C. *et al.*, The carbon cycle and atmospheric carbon dioxide , Chapter 3, in [8], 2001.

[114] P. Friedlingstein, I. Fing, E. Holland, *et al.*, On the contribution of CO_2 fertilization to the missing biospheric sink. *Global Biogeochemical Cycles*, **9** (1995), 541–56.

[115] T. H. Jones, L. J. Thompson and J. H. Lawton, Impacts of rising atmospheric carbon dioxide on model terrestrial ecosystems. *Science*, **280** (1988), 441–3.

[116] R. J. Norby, E. H. DeLucia, B. Gielen, *et al.*, Forest response to elevated CO_2 is conserved across a broad range of productivity. *Proceeding of the National Academy of Sciences of the USA*, **102**:n°50 (2005), 18052–6.

[117] B. A. Hungate, J. S. Dukes, M. R. Shaw, Y. Luo and C. B. Field, Nitrogen and climate change. *Science*, **302** (2003), 1512–3.

[118] R. R. Nemani, C. D. Keeling, H. Hashimoto, *et al.*, Climate-driven increases in global terrestrial Net Primary Production from 1982 to 1999. *Science*, **300** (2003), 1560–3.

[119] P. M. Cox, R. A. Betts, C. D. Jones, S. A. Spall and I. J. Totterdell, Acceleration of global warming due to carbon-cycle feedbacks in a coupled climate model. *Nature*, **408** (2000), 184–7.

[120] M. Berthelot, P. Friedlingstein, P. Ciais, et al., Global response of the terrestrial biosphere to CO_2 and climate change using a coupled climate-carbon cycle model. *Global Biogeochemical Cycles*, **16** (2002).

[121] W. Cramer, A. Bondeau and F. I. Woodward, Global response of terrestrial ecosystem structure and function to CO_2 and climate change: results from six dynamic global vegetation models. *Global Change Biology*, **7** (2001), 357–73.

[122] M. Cao and F. I. Woodward, Dynamic responses of terrestrial ecosystem carbon cycling to global climate change. *Nature*, **393** (1998), 249–52.

[123] J.-L. Dufresne, P. Friedlingstein, M. Berthelot, et al., On the magnitude of positive feedback between future climate change and the carbon cycle. *Geophysical Research Letters*, **29** (2002), 1–4.

[124] P. Friedlingstein, et al., Climate-carbon cycle feedback analysis; results from the C4MIP model intercomparison. *Journal of Climate* (2006), 3337–53.

[125] L. Bopp, L. Legendre and P. Monfray, La pompe à carbone va-t-elle se gripper? *La Recherche*, **355** (2002), 49–51.

[126] F. Joos, R. Meyer, M. Bruno and M. Leuenberger, The variability in the carbon sinks as reconstructed for the last 1000 years. *Geophysical Research Letters*, **26** (1999), 1437–40.

[127] J. L. Sarmiento and C. Le Quere, Oceanic carbon dioxide uptake in a model of century scale global warming. *Science*, **274** (1996), 1346–50.

[128] T. Blunier, J. A. Chappellaz, J. Schwander, et al., Atmospheric methane, record of a Greenland ice core over the last 1,000 years. *Geophysical Research Letters*, **20** (1993), 2219–22.

[129] *ACIA Overview Report: Impacts of a Warming Arctic* (Cambridge: Cambridge University Press, 2004). (pdf files available on www.acia.uaf.edu)

[130] *ACIA Scientific Report: Arctic Climate Impact Assessment* (Cambridge: Cambridge University Press, 2005). (pdf files available on www.acia.uaf.edu)

[131] *ACIA Policy Report: Arctic Climate Impact Assessment: Policy Document* (Cambridge: Cambridge University Press, 2004). (pdf files available on www.acia.uaf.edu)

[132] *Millennium Ecosystem Assessment Global Assessment Reports*, vol. 1: current state & trends (Washington, DC: Island Press, 2005). (pdf files available on www.millenniumassessment.org/en/products.aspx)

[133] J. Räisänen, CO_2-induced changes in interannual temperature and precipitation variability in 19 CMIP2 experiments. *Journal of Climate*, **15** (2002), 2395–411.

[134] IPCC /TEAP, Special report on safeguarding the ozone layer and the global climate system: Issues related to hydrofluorocarbons and perfluorocarbons. Prepared by the W.G. I and II of IPCC and the Technology and Economic Assessment Panel (United Kingdom/New York: Cambridge University Press, 2005).

[135] V. Ramaswamy, M.-L. Chanin, J. Angell, et al., Stratospheric temperature trends: observations and model simulations. *Reviews of Geophysics*, **39** (2001), 71–122.

[136] European Commission, Ozone-climate interactions, in *Air Pollution Research Report*, ed. I. S. A. Isaksen, Report 81, EUR 20623, Luxemburg, 2003.

[137] WMO-UNEP Scientific Assessment of Ozone Depletion, Global Ozone Research and Monitoring Project – Report N°50, Geneva, Switzerland, 2007.

[138] US National Research Council, Board on Atmospheric Sciences and Climate, Commission on Geosciences, Environment, and Resources, *From Research to Operations in Weather*

Satellites and Numerical Weather Prediction: Crossing the Valley of Death (Washington, DC: National Academy Press, 2000).

[139] GCOS Second Adequacy Report, The second report on the adequacy of the global observing systems for climate in support of the UNFCCC. GCOS-82 (WMO/TD n°1143). 2003. (pdf available on http://www.wmo.ch/web/gcos/gcoshome.html)

[140] GCOS Implementation Plan, Implementation plan for the global observing system for climate in support of the UNFCCC. GCOS-92 (WMO/TD n°1219). 2004. (pdf available on http://www.wmo.ch/web/gcos/gcoshome.html)

[141] GCOS Satellite Supplement, Systematic observation requirements for satellite-based products for climate. GCOS-107 (WMO/TD n°1338). 2006. (pdf available on http://www.wmo.ch/web/gcos/gcoshome.html)

[142] CEOS Response to GCOS Requirements, Satellite observation of the climate system. 2006. (pdf available at http://www.ceos.org/pages/pub.html)

[143] M. Molina and F. S. Rowland, Stratospheric sink for chlorofluoromethanes: chlorine atomic catalyzed destruction of ozone. *Nature*, **249** (1974), 810–812.

[144] R. S. Stolarski and R. J. Cicerone, Stratospheric chlorine: a possible sink for ozone. *Canadian Journal of Chemistry*, **52** (1974), 1610–15.

[145] National Research Council, *Halocarbons: Effects on Stratospheric Ozone*, (Washington, DC, 1976).

[146] R. E. Benedict, *Ozone Diplomacy: New Directions in Safeguarding the Planet* (Cambridge: Harvard University Press, 1991).

[147] H. L. Bryden, H. R. Longworth and S. A. Cunningham, Slowing of the Atlantic meridional overturning circulation at 25°N. *Nature*, **438** (2005), 655–7.

[148] B. Tissot, *Halte au Changement Climatique*, (Odile Jacob, Sciences, 2003).

[149] R. Socolow, C. Andrews, F. Berkhout and V. Thomas, *Industrial Ecology and Global Change* (New York: Cambridge University Press, 1994).

Index

aerosols, 7,16,19, 20, 31, 33, 38, 39, 50, 52, 53, 54, 55, 56, 57, 58, 59, 60, 61, 177, 183, 185, 194, 197, 218, 219, 220, 223, 224, 233, 238, 239
　absorbing, 53
　amounts, 17
　anthropogenic, 16, 52, 53
　atmospheric, 16, 49, 55, 57, 224
　concentration, 50, 54, 57
　effect, 17, 39, 48, 52, 54, 55, 58, 233
　emission, 7, 59
　forcing, 60
　geoengineering, 60
　measurements, 56, 57
　natural, 54
　observations, 52
　oceanic (also salt), 51, 52, 60
　optical thickness, 51, 57
　organic, 181
　parameters, 48
　particle, 54
　pollutants, 16
　precursor, 59
　primary, 50
　production, 44
　properties, 56, 57, 58
　remote sensing, 57
　role, 25
　satellite retrievals, 56
　scattering, 56
　secondary, 50
　smoke, 54
　soot, 220
　stratospheric, 17, 60
　sulfate, 1, 50, 59, 179
　tropospheric, 52
　types, 49, 50
　volcanic, 19, 31, 224
　See also specific pollutants
Antarctic, 12, 14, 19, 23, 110, 112, 113, 115, 116, 117, 118, 120, 179, 182, 184, 185, 238, 246
　cooling, 117
　ice caps, 72, 101
　ice core, 123, 125
　ice extent, 113
　ice mass, 113,
　ice sheet, 8, 21, 105, 115, 116, 118, 120
　oscillation, 117
　ozone depletion (also ozone hole), 117, 179
　Peninsula, 113, 117
　surface temperature, 117
　West, 115, 119
Arctic, 1, 8, 21, 70, 77, 81, 89, 115, 117, 118, 119, 149, 151, 152, 153, 154, 155, 157, 181, 184, 233, 237, 238
　agriculture, 153
　air temperature, 118, 156
　basin, 152, 154
　climate, 154
　climate impact assessment (ACIA), 119, 149, 151, 234
　council, 151, 234
　feedback, 118
　forcing, 118
　ice, 70
　indigenous people, 154
　ocean, 70, 79, 103, 113, 117, 221
　oscillation, 118
　ozone, 184
　research, 237
　sea ice, 8, 21, 105, 108, 111, 113, 114, 118, 119, 152, 221, 245
　system, 114, 245
　temperature, 21
　tundra, 244
　vegetation zones, 154
　warming, 151, 154

253

Carbon cycle, 8, 40, 91, 121, 123, 124, 128, 130, 131, 132, 133, 134, 135, 138, 139, 142, 144, 178, 183, 200, 218
  fast, 124
  geological, 124, 135
  global, 125, 127, 128, 129, 131, 132, 133, 134, 135, 137, 139, 141, 143, 145, 147, 224
  interglacial, 130
  natural, 138
  ocean, 129, 130, 142
  terrestrial, 130, 132
Carbon dioxide, 14, 15, 16, 17, 24, 36, 40, 41, 63, 64, 87, 88, 95, 122, 144, 173, 176, 177, 178, 194, 205, 234, 236, 246, 247, 248, 249, 250
  concentration, 14, 20, 23, 68, 128, 144, 163, 197
  emissions, 15, 136, 151, 177, 228
  measurements, 14
  uptake, 108, 129
Chlorofluorocarbons, 15, 23, 40, 177, 178, 183, 209, 213, 219, 234, 235
  residence time, 179, 183
  See also specific CFCs
Clouds, 3, 5, 7, 16, 19, 31, 33, 34, 38, 39, 40, 43, 45, 46, 47, 48, 49, 53, 54, 55, 58, 61, 63, 66, 70, 71, 72, 85, 170, 193, 197, 218
  altitude, 39, 40, 46
  cirrus, 60, 66, 181
  climatology, 32
  condensation nuclei (CCN), 38, 54, 55
  convective, 35, 63, 70, 170
  cover, 118, 220, 225
  crystal, 39
  cumulonimbus, 67
  droplet, 38, 39, 49, 54, 55, 220
  dynamics, 67
  effects, 7, 35, 44, 45, 94
  feedback, 46, 48
  formation, 44, 45, 53, 54, 86
  lifetime, 44
  microphysics, 45, 46, 66
  model, 73
  observations, 32, 38
  particles (also droplets), 16, 66, 68, 186
  phase, 46
  polar stratospheric, 179
  radiation interaction, 19, 45, 223
  radiative forcing, 38
  representation (*see also* Model), 45
  stratus, 66, 70
  thickness, 38, 39
  types, 32
  water content, 46, 54, 58
Coal, 10, 37, 135, 136, 137, 226, 227, 228
  clean, 226
  combustion (also burning), 181, 226
  mining, 138, 181
  use, 137
Conveyor belt, 102, 128, 129, 143
  See also specific thermohaline circulation
Cryosphere, 5, 8, 93, 110, 112, 117, 120

Deforestation, 4, 7, 15, 50, 80, 88, 131, 133, 134, 135, 173, 220
  consequence, 87
  effect, 88
El Niño, 14, 41, 99, 134, 142, 174, 230
  conditions, 100
  dynamics, 100
  event, 5, 174, 221
  occurrence, 21
  year, 13, 51, 220
Evaporation, 43, 55, 63, 66, 71, 75, 76, 79, 80, 83, 84, 85, 86, 87, 89, 104, 170, 235
  flux, 53
  process, 45
  rate, 60, 79

Feedback, 5, 6, 7, 19, 31, 42, 43, 44, 45, 46, 48, 55, 60, 70, 80, 85, 88, 111, 114, 117, 118, 119, 127, 130, 131, 141, 142, 143, 175, 177, 218, 220, 234
  atmospheric, 43
  biological, 142
  chemical, 141, 185
  cloud, 46, 48
  effect, 42, 44
  land-climate, 87
  negative, 45, 185
  polar, 111
  positive, 8, 45, 79, 114, 117, 118, 120, 142, 144, 158, 185
  snow albedo, 44, 53, 114, 117
  thermohaline, 117
  vegetation, 44
  water vapor, 43
Floods, 78, 84, 99, 102, 170, 223, 230
  centennial, 170
  spring, 167
  winter, 169

Index   255

Forcing, 6, 14, 16, 18, 21, 33, 69, 104, 112, 117, 118, 158, 161, 177, 183, 192, 206, 234, 235
  aerosol, 60
  anthropogenic, 18, 19, 48
  dynamic, 30
  greenhouse, 44, 54, 81, 144
  natural, 16, 17, 18, 239
  radiative, 33, 34, 35, 37, 38, 39, 40, 44, 52, 53, 54, 60, 138, 181, 184, 185, 220, 224
  solar, 150
  volcanic, 150

Glacier, 2, 7, 70, 76, 77, 78, 107, 110, 111, 112, 115, 199, 120, 150, 151, 152, 157, 192, 194, 195, 236
  continental, 11, 112, 113, 225
  Greenland, 9
  marine, 113
  mountain, 8, 105, 119, 221
  outlet, 112, 115, 118, 120, 238
  polar, 118
Greenhouse effect, 7, 13, 14, 14, 16, 17, 23, 24, 25, 26, 32, 36, 39, 43, 63, 64, 66, 183, 220, 236
  anthropogenic, 37
  cloud, 39
  concept, 35
  enhancement, 181
  natural, 42, 127
  role, 45
Greenhouse gases, 5, 15, 19, 31, 33, 34, 36, 37, 38, 40, 42, 43, 52, 59, 64, 78, 80, 86, 88, 95, 117, 122, 127, 128, 150, 157, 177, 178, 180, 184, 205, 224, 234, 239
  concentration, 24, 37, 48, 144, 185, 220
  effects, 16
  emissions, 16, 18, 20, 228
  human-caused (also anthropogenic), 10, 19, 60, 144, 219
  production, 205, 212
Greenland, 8, 14, 21, 101, 115, 152, 182, 206, 234
  glaciers, 9
  ice caps, 72
  ice sheet, 19, 105, 111, 112, 115, 116, 118, 119, 120, 152, 218, 221
  water, 154

Hadley circulation, 35, 44, 66, 236
Hurricane, 3, 5, 63, 79, 86, 87, 170, 218, 230
  Atlantic, 21
  Katrina, 10

Ice cores, 182, 222
IPCC, 2, 5, 16, 19, 20, 22, 23, 24, 25, 26, 27, 29, 40, 43, 68, 70, 77, 80, 124, 133, 136, 138, 150, 151, 152, 156, 158, 162, 191, 213, 214, 215, 234, 236, 243, 250
Isotopes, 15
  carbon, 126, 133
  radioactive, 227

Kyoto Protocol, 2, 9, 10, 23, 25, 26, 136, 145, 157, 215, 216, 228, 237

Lidar, 46, 185, 194

Mesosphere, 185, 237
Methane, 15, 36, 123, 124, 158, 181
Model, 6, 7, 18, 19, 25, 32, 44, 45, 48, 64, 67, 68, 71, 80, 82, 85, 86, 88, 102, 103, 104, 117, 118, 131, 142, 161, 162, 163, 164, 167, 168, 170, 171, 172, 178, 184, 185, 187, 192, 197, 200, 210, 223, 226
  atmospheric, 78, 104, 105
  biogeochemical, 131
  biological, 141
  carbon-climate, 138
  climate, 3, 6, 7, 17, 18, 19, 21, 22, 25, 42, 43, 44, 45, 52, 65, 70, 80, 104, 118, 140, 150, 161, 162, 163, 164, 171, 177, 179, 198, 212, 225, 230, 236
  cloud, 46, 72
  computer (also numerical), 3, 24, 129, 130, 131
  consensus, 80, 88
  coupled, 140
  Earth System, 195
  ecological, 131
  general circulation, 17, 71, 103, 145, 225
  global, 46, 142
  ice, 104, 113, 119
  numerical, 6, 8, 18, 100, 103, 104, 105, 108, 161, 191
  ocean, 141
  physiological, 131
  projection, 3, 18, 70, 105, 221, 223
  radiative transfer, 32
  resolution, 48, 67
  result, 20
  simulation, 6, 23, 46, 68, 89, 138, 152

Model (*cont.*)
  terrestrial carbon, 140
  uncertainties, 42, 43
  water cycle, 73

Observing system, 9, 100, 192, 196, 197, 198, 199, 200
  carbon, 146
  climate, 191, 192, 196, 197, 200, 2002
  Earth, 200, 201, 202
  future, 196
  global, 190, 196, 197, 198
  in situ, 108, 196
  ocean, 99, 101
  operational, 225
Ozone, 36, 40, 64, 177, 179, 182, 184, 185, 187, 194, 205, 234, 238
  concentration, 119, 179, 183, 184, 185, 186, 219
  depleting, 184, 188
  depletion, 41, 117, 178, 179, 184, 187, 188, 213
  destruction, 183
  effect, 182
  formation, 180
  molecule, 179, 180, 181, 183
  precursor, 180, 186
  radiative forcing, 41
  recovery, 184, 186
  stratospheric, 23, 178, 179, 181, 182, 184, 185, 186, 187, 188, 208, 209, 213, 219, 220, 239
  surface, 180, 183
  tropospheric, 41, 180, 182, 186, 219, 220

Paleo
  climatic records, 3, 118, 142
  climatic studies, 116, 119

climatic timescale, 44
climatologist, 25, 102, 190
climatology, 224
  oceanography, 101
Permafrost, 75, 78, 89, 140, 151, 153, 181, 238
Photosynthesis, 15, 130, 131, 140, 176, 229
Precipitation, 8, 14, 16, 21, 22, 39, 45, 53, 54, 55, 58, 61, 63, 70, 71, 73, 76, 77, 78, 79, 80, 81, 82, 84, 86, 88, 89, 90, 104, 127, 162, 167, 168, 173, 217, 221, 239
  amount, 3, 55, 174, 177, 186, 194, 195
  annual, 81
  change, 14, 164
  convective, 71
  development, 54
  distribution, 55, 221
  efficiency, 66
  event, 89, 90
  forecast, 71, 72
  formation, 54
  intense, 86, 164, 170
  liquid, 167, 234
  mean, 163
  measurement, 164
  process, 16, 54, 60, 72
  record, 55
  regime, 169, 172
  snow, 112, 115, 116
  solid, 83
  summer, 169
  variability, 79
  winter, 169

Radar, 46, 113, 115, 193, 197, 198, 233
Radiometer, 32, 197, 237, 238, 240

Salinity, 63, 71, 92, 93, 96, 99, 100, 102, 104, 107, 111, 195, 197, 225
Satellites, 13, 33, 43, 52, 58, 100, 101, 113, 171, 192, 193, 195, 202, 221, 234
  altimetric, 106, 107, 108
  artificial, 93
  constellation, 194, 197
  Earth-observing, 139, 200
  environmental, 196, 198
  geostationary (also geosynchroneous), 32, 57
  observations, 52
  orbiting, 73
  remote sensing, 193, 239
  weather, 193
Sea ice, 2, 8, 22, 110, 111, 112, 113, 117, 150, 152, 155, 194, 225, 235, 236
  albedo, 114
  Antarctic, 111
  Arctic, 8, 21, 105, 108, 111, 113, 114, 118, 152, 221
  coverage, 111, 117
  extent, 20, 112, 114, 117, 118, 119, 152, 220, 225, 239
  glacier, 157
  model, 113
  monitoring, 196
  observation, 113
  pack, 114
  reduction, 120
  seasonal, 111, 113
  summer, 152
  thickness, 113, 118, 151
  winter, 113
Sea level, 1, 9, 14, 17, 21, 22, 106, 115, 116, 151, 152, 153, 192, 194, 221
  change, 105, 106, 107, 240

global, 106, 199
mean, 7, 101, 105, 108
measurement, 202, 225
rise, 8, 14, 21, 22, 24, 70, 105, 107, 108, 112, 115, 116, 117, 118, 119, 120, 154, 206, 221, 224, 230
signal, 107
Solar energy, 30, 31, 36, 37, 53, 60, 88, 89, 114, 121, 227, 233
Storms, 63, 66, 71, 77, 79, 86, 155
Stratosphere, 17, 33, 40, 41, 60, 178, 179, 180, 181, 182, 183, 184, 185, 186, 187, 224, 239, 240

Troposphere, 7, 33, 39, 40, 41, 43, 178, 179, 180, 181, 182, 187, 235, 239, 240
top, 68, 220, 235
upper, 35, 44, 64, 182, 186, 234, 236
warming, 185

UNFCCC, 2, 24, 25, 157, 184, 203, 240, 250

Water cycle, 5, 7, 53, 60, 61, 75, 76, 78, 80, 84, 85, 89, 131, 200
atmospheric, 63, 88, 169
global, 71, 72, 95
terrestrial (also continental), 77, 80, 82, 83, 88
model, 73
variable, 80
See also specific hydrological cycles
Water vapor, 7, 34, 36, 37, 43, 44, 54, 55, 58, 63, 64, 65, 75, 77, 85, 86, 87, 89, 131, 177, 182, 194, 221, 225, 234, 235, 239
anthropogenic, 40
atmospheric, 38, 63, 64, 76, 87, 88, 221
budget, 64
column, 57

concentration, 43
condensation, 66, 71, 89, 170
distribution, 71
feedback, 43, 44, 46
molecule, 39, 182
natural, 41
oxidation, 177, 178
stratospheric, 41
transport, 71
tropospheric, 44, 64
Weather, 63, 66, 73, 86, 88, 92, 119, 153, 187, 191, 193, 199, 207, 223, 237, 240
fluctuation, 69
pattern, 70, 120, 131, 186
phenomena, 67
prediction (also forecast), 71, 99, 189, 196, 230
satellite, 193
station, 13
system (also disturbance), 34, 63, 66, 67, 69, 70, 71, 73, 235